Energy
Engine of Evolution

Cover illustration

X-ray image of Sun: courtesy Lockheed Solar and Astrophysics Laboratory, Palo Alto, Calif. Originally published in Projects in Scientific Computing, 1994, © Pittsburgh Supercomputing Center, 1994.

Energy
Engine of Evolution

Frank Niele

Shell Global Solutions

2005

ELSEVIER

Amsterdam · Boston · Heidelberg · London · New York · Oxford · Paris
San Diego · San Francisco · Singapore · Sydney · Tokyo

ELSEVIER B.V.　　　　　　ELSEVIER Inc.　　　　　ELSEVIER Ltd　　　　　　　　ELSEVIER Ltd
Radarweg 29　　　　　　　525 B Street　　　　　　The Boulevard　　　　　　　 84 Theobalds Road
P.O. Box 211, 1000 AE　　　Suite 1900, San Diego　 Langford Lane, Kidlington,　London WC1X 8RR
Amsterdam, The Netherlands　CA 92101-4495, USA　　Oxford OX5 1GB, UK　　　 UK

© 2005 Sheel Global Solutions International BV. Published by Elsevier BV. All rights reserved.

This work is protected under copyright by Shell Global Solutions International BV, and the following terms and conditions apply to its use:

Photocopying
Single photocopies of single chapters may be made for personal use as allowed by national copyright laws. Permission of the Publisher and payment of a fee is required for all other photocopying, including multiple or systematic copying, copying for advertising or promotional purposes, resale, and all forms of document delivery. Special rates are available for educational institutions that wish to make photocopies for non-profit educational classroom use.

Permissions may be sought directly from Elsevier's Rights Department in Oxford, UK: phone: (+44) 1865 843830, fax: (+44) 1865 853333, e-mail: permissions@elsevier.com. Requests may also be completed on-line via the Elsevier homepage (http://www.elsevier.com/locate/permissions).

In the USA, users may clear permissions and make payments through the Copyright Clearance Center, Inc., 222 Rosewood Drive, Danvers, MA 01923, USA; phone: (+1) (978) 7508400, fax: (+1) (978) 7504744, and in the UK through the Copyright Licensing Agency Rapid Clearance Service (CLARCS), 90 Tottenham Court Road, London W1P 0LP, UK; phone: (+44) 20 7631 5555; fax: (+44) 20 7631 5500. Other countries may have a local reprographic rights agency for payments.

Derivative Works
Tables of contents may be reproduced for internal circulation, but permission of the Publisher is required for external resale or distribution of such material. Permission of the Publisher is required for all other derivative works, including compilations and translations.

Electronic Storage or Usage
Permission of the Publisher is required to store or use electronically any material contained in this work, including any chapter or part of a chapter.

Except as outlined above, no part of this work may be reproduced, stored in a retrieval system or transmitted in any form or by any means, electronic, mechanical, photocopying, recording or otherwise, without prior written permission of the Publisher. Address permissions requests to: Elsevier's Rights Department, at the fax and e-mail addresses noted above.

Notice
No responsibility is assumed by the Publisher for any injury and/or damage to persons or property as a matter of products liability, negligence or otherwise, or from any use or operation of any methods, products, instructions or ideas contained in the material herein. Because of rapid advances in the medical sciences, in particular, independent verification of diagnoses and drug dosages should be made.

First edition 2005
Reprinted 2006

British Library Cataloguing in Publication Data
A catalogue record is available from the British Library.

Library of Congress Cataloging in Publication Data
A catalog record is available from the Library of Congress.

ISBN-13: 978 0 444 51886 6
ISBN-10: 0 444 51886 X

Working together to grow
libraries in developing countries

www.elsevier.com | www.bookaid.org | www.sabre.org

ELSEVIER　　BOOK AID International　　Sabre Foundation

∞ The paper used in this publication meets the requirements of ANSI/NISO Z39.48-1992 (Permanence of Paper).
Printed in The Netherlands

Contents

About the book ix

Foreword xi

Acknowledgements xiii

Introduction xvii

PART I   THE FIRST FIVE ENERGY REVOLUTIONS
— a time journey through the history of life —

Chapter 1   The Photo-Energy Revolution

The origin of life	3
The origin of sunlight	3
The origin of planet Earth	3
The origin of the living cell	5
The Thermophilic Regime	8
Photosynthesis and free oxygen	8
The origin of photosynthesis	8
Sunlight-driven energy technology	9
The Phototrophic Regime	10
And the face of the earth changed	11
Two grand carbon cycles	11
The self-inflicted crisis	12

Chapter 2   The Oxo-Energy Revolution

The origin of aerobic respiration	13
The origin of breathable air	13
Global oxygen crisis	14
Aerobic respiration	15
Oxygen-driven energy technology	16
Microbial symbiosis	18
The origin of the eukaryotic cell design	18
The microbial energy majors	20

The Aerobic Regime	21
Biological diversification	22
Ecological growth	24
And the face of the earth changed	**25**
The changing landscape	25
Competition and adaptation	26

Chapter 3 The Pyro-Energy Revolution

The origin of hominids	**29**
It needed a crisis	29
The origin of human culture	30
The quest for the human advantage	32
Man the fire master	**38**
The human advantage	38
The origin of fire	41
Active use of fire	41
The Pyrocultural Regime	43
Societal metabolism	45
The Symbolisational Signal	46
And the face of the earth changed	**48**
A changing lifestyle	48
And life changed the land	50

Chapter 4 The Agro-Energy Revolution

Man the solar farmer	**51**
The Agricultural Revolution	51
The forces that shaped the revolution	52
The Agrocultural Regime	54
The emergence of competing energy chains	55
A new reality emerges	**56**
The Quantificational Signal	56
The Scientific Revolution	57
And the face of the earth changed	**59**
From farms to nation states	59
The growing human footprint	61

Chapter 5 The Carbo-Energy Revolution

Hydrocarbon Man	**65**
The agricultural energy crisis	65
The dawn of ancient sunlight	66
The Carbocultural Regime	70

Contents

The Carbian Explosion	71
An explosion of competing energy chains	71
Electric energy as information carrier	74
Carbocultural non-fossil energy chains	76
Sensitive scientific senses	79
And the face of the earth changed	81
The surging human footprint	81
From fire master to fire addict	83

PART II   THE NEXT ENERGY REVOLUTION
— evolutionary energetics, models and scenarios —

Chapter 6   The Staircase of Energy Regimes

Energy regimes	87
The foundation of the Staircase model	87
The rise and fall of complexity	98
The evolution of information	101
Generating information	101
From genes to memes and artefacts	103
Socio-technological development	106
The rhythm of societal development	106
The pushes and pulls of progress	113
The evolution of system earth	114

Chapter 7   The Emerging Helio-Energy Revolution

Signs of a coming energy revolution	117
The Macroscopical Signal	117
Two perceptions of reality	119
Leaving Carbon Valley	121
Nuclear Valley	122
Green Valley	125
Sustainable development	128
Limits to nuclear	131
Limits to green	132
The next energy revolution	136
Symbian Man	136
The Helio-Energy Revolution	139
Sun Valley	143
Socio-metabolic complexes	145
And the face of the earth will change	147

Appendix Energy, Complexity and Evolution
    The nature of energy 149
    Energy-driven organisation 153

References 163

Glossary 171

Index 183

About the book

Energy is an important topic for Shell Global Solutions as the organisation is a major provider of services to the oil and gas industry. This book describes the role that energy has played in the evolution of nature and culture on our planet. It provides a compact history of energy over the last four billion years, with the aim of creating a sound basis for understanding the possible futures of the energy industry.

Frank Niele describes with great insight the results of years of detailed study and discussions with his colleagues in our exploratory research laboratories and in the academic world.

Last year I was pleased to introduce the book Insight in Innovation written by Jan Verloop and it is very gratifying that Shell Global Solutions is again in a position to sponsor an important book written by one of its employees. These two books could well become the start of a limited series of 'thought-leadership' books on issues that are relevant for the industry we service. With innovation and energy as the first two topics, we are off to a promising start.

Greg Lewin
President Shell Global Solutions

Foreword

The number of books written about energy, and the impact that energy has on all aspects of our lives, must be in the thousands. But this compelling book tells a unique and different story. It is a story that starts four billion years ago with the origin of life on earth, and that ends a hundred years into the future where a picture unfolds of a new energy world.

Frank Niele's book describes the relationship between life and energy through time, and outlines how the major revolutions in the evolution of life on earth were driven by developments at the energy frontiers. Humankind started his special relationship with energy 500,000 years ago when he first learned how to control fire. Over time, people have created increasingly bigger and more complex energy systems to support their way of life. But this progress has been accompanied by a growing concern about the impact of the 'human footprint on our planet' and the way we use our natural resources. The message in this book is that we are on the verge of the next energy revolution, where we will learn how to master new energy forms in a new way.

The book also brings a fresh element into the discussion of sustainable development. Energy is one of the essential components in sustainable development, as without an ample supply of clean and affordable energy, the world cannot successfully address the important equity and ecology issues that it faces. An outlook on a sustainable energy future that has a sound scientific basis and learns from the lessons of five earlier energy revolutions is a timely and valuable contribution to the sustainable development debate. Frank's book provides a positive perspective on what is in many ways an uncertain energy future.

This book should appeal to a readership that is much wider than just the energy industry. Anyone interested in a 'sustainable future of system Earth' should find the ideas and insights in this book stimulating. But I have to give a minor word of warning. The issues that the world faces in the area of clean and reliable supply of energy are not simple and this book does not give its answers away without some effort from the reader – an effort that will be well rewarded though as the story unfolds.

Rob Routs
Executive Director Downstream Shell International

Acknowledgements

In the second half of the 1990s the concept of 'sustainable development' entered the world of the large corporations, at the same time as 'shareholder value' became the leading performance target. The grand paradox was that 'sustainable development' demands a longer-term view, while 'shareholder value' needs a short-term focus. Against this tide, Michiel Groeneveld, then manager of Exploratory Research in Shell Global Solutions, had both the vision and the courage to explore the long-term relationships between sustainable development and energy, and created space for me to contribute. We did not have the slightest idea of what we had embarked upon: an in-depth study that ended with this book on the evolution of energy species. Michiel inspired me to create, and, since that is perhaps the most special award a scientist can receive, I profoundly thank him for that, for reviewing my manuscripts, and for the many fine discussions we have had.

This book could only come into being in a fertile mental environment such as that offered by the Exploratory Research group of Shell Global Solutions, home to an international cluster of scientists, technologists, business analysts and technicians with a remarkable palette of skills and a source of truly non-conformist thinking. I thank all my colleagues, in particular Wim Wieldraaijer, Evert Wesker, and Jan Werner for the often stirring dialogues that continuously shaped and sharpened my mind.

I am very much indebted to Hans Gosselink, my manager, who shared all the ups and downs from the beginning to the end with total support. After reviewing each draft Hans always provided the most valuable comments and suggestions. Because of his kind but persistent challenging, I went quite deeply into the evolution of culture, to end up with its energy-driven beginning long before the origin of human civilisation.

I am grateful to Gert Jan Kramer for his critical reviews of my drafts. Gert Jan's aid was invaluable. Each inconsistency he spotted, whether physical, historical, or whatever, was followed by a most enjoyable discussion. It was he who recommended Donald Worster's 'Nature's Economy', a work essentially instrumental to me in collecting my thoughts. I also thank Hans Geerlings for reviewing my pre-book scribbles and exploring the borders between scientific disciplines. Hans moves most easily from physics to philosophy or memetics. But above all, he helped me to cross the limits of conventional thermodynamics in an attempt to describe the evolutionary energetics of life, both biological and cultural. The last colleague whom I should like to thank is Andreas Nowak. Together we ran a project aimed at turning part of the theory into practice and that resulted in shaping the so-called CityPlex concept. Refinery experts capable of applying their knowledge and know-how to entirely different metabolic schemes are both rare and crucial to forward sustainable development. Having had the chance to work with Andreas was nothing less than a privilege.

My final manuscript went through two intermediate drafts. The first was read by several senior people from inside and outside Shell. I thank Doug McKay, who in those days led the Shell long-term energy scenarios team, for his valuable support and advice. I remember our dialogues with pleasure. I also thank Kurt Hoffman, Director of the Shell Foundation. Kurt agreed with the way my story links energy and sustainable development. The same holds for André Smit, Manager of Social Investment, to whom I am indebted for his enthusiastic engagement. From the world outside Shell, Professors Wim van Swaaij and J. Goudsblom strongly encouraged me to continue my research for which I was, and still am, grateful. Support from such eminent scholars gives one that extra stimulus to go on.

I once asked my colleague and fellow chemist Rob van Veen, a widely read man, about the emergence of modern science. Rob recommended I read 'The Measure of Reality' by historian Alfred Crosby, a pivotal work that influenced my thinking enormously, and is therefore worth an expression of gratitude. I also thank Emile van Kreveld, an esteemed, now retired, colleague who still frequently visits the laboratory, for the many wonderful and instructive conversations we have had. Emile has this special talent to grasp the essence of natural as well as cultural phenomena and suggested several valuable readings to me.

When the second draft was ready we invited two authorities in the sustainable development sciences, Professor Jeroen van den Bergh and Dr. Bob van der Zwaan, to review this version. Both recommended that we seek a premier publisher and suggested that I should incorporate more on the economics of sustainable development. But also, completely independently from each other, both advised reading E.O. Wilson's 'Consilience: The Unity of Knowledge'. And indeed, not integrating Wilson's work would have been a sin of omission. I thank Jeroen and Bob for their supportive and critical advice.

I would like to thank also our Group Research Advisor, Peter Kwant. Together with Michiel, Peter enabled the birth of this work and additionally helped it reach ripeness in cooperation with Michiel's successor Herman van Wechem. Thanks to Herman, for so large-heartedly backing

Acknowledgements

me in finishing this major, unconventional effort. I am grateful to Greg Lewin, President of Shell Global Solutions, not only for enabling me to publish this book, but also for his commitment to supporting an academic work.

Because of the dual role he played, I purposely omitted to mention one man in the acknowledgements listed above: Jan Verloop. In his capacity as Innovation Manager, Jan gave me the opportunity to carry out long-term exploratory projects and always was supportive. Then, after his retirement, he responded positively to a request to help me as structural editor in converting the second draft into a readable manuscript. Communicating science in almost-normal language is far from easy. If I did succeed – and that is for you, the reader, to judge – then appreciate Jan's contribution. But also, because of Jan's help, this book has become a true monograph about evolutionary energetics, without me dwelling at length on all those other aspects of life and energy that fascinate me. A special word of thanks therefore to Jan for his numerous constructive comments and suggestions, both about the broad approach as well as about the smallest details, and his patient and enjoyable cooperation. Then I should also like to thank Gill Rosson who superbly edited the final draft. Gill has this unique gift to edit a sentence in such a way that the writer wished he had written it and thus thinks it is his.

My final words of deep thanks go to Cora. During our many walks through dunes and over dikes, we fused natural and cultural perspectives on life in attempts to understand the idea of sustainable development. Cora reviewed all first drafts before they left the house, so helping me write clearly and coherently. Though I know she will look back with mixed feelings to the origin of this book, in fact the escalation of an idea – it could not have been done without her unfailing help and support.

Introduction

This book is an unplanned side effect of my struggle to understand the conflict between society's growing demand for economical energy and the quest for sustainable development. The root cause of this conflict is the growing paradox that utilisation of fossil fuels energises economic growth while simultaneously inducing local pollution and global climate change; it thus creates and affects prosperity at the same time, albeit that today's balance looks good while tomorrow's is questionable.

This contentious issue became urgent for energy companies in the late 1990s. At the time I was working in the Exploratory Research group of Shell Global Solutions where researchers study technological options with a view to creating new business opportunities. The primary goal has always been to convert crude oil and natural gas into quality fuels and petrochemical products. But the call for sustainable development challenged the logic of this approach and forced us to reconsider the many, often implicitly selected choices that were being made. Indeed, it became imperative to innovate energy chains from source to service to sink.

Michiel Groeneveld, my boss, created the space for me to distance myself from day-to-day worries and to explore the dynamics of the energy economy as a whole in an attempt to discover possible relationships between 'energy' and 'sustainable development'. What came out of the study is a systemic description of the role of energy in *life*, for focusing on human societies only made no sense. This book aims to describe my findings in the form of a scientific monograph.

The monograph consists of two parts. Part I, the first five chapters, describes a journey through the history of planet Earth along the tracks left by energy. I have melded numerous observations and reflections made by scientists from many different disciplines into a logical narrative. These five chapters correspond with the five successive energy revolutions I discovered during my travel through time. An *energy revolution* marks the boundaries between *energy periods*, and an energy period is characterised by an ecologically dominant *energy regime*. Earth's palaeo-record reveals how each energy revolution induced the emergence of radically new origins with corresponding new developments, and how, with each revolution, the face of the Earth changed drastically.

Part II comprises two chapters and an appendix. The first chapter describes a model for the role of energy in the evolution of life: this model is called the Staircase of Energy Regimes. The second chapter applies the Staircase model in exploring pathways to how I envision a more sustainable future, while the appendix provides a theoretical foundation for the Staircase model based on an analysis of the observations reported in Part I. The appendix is written in plain English, without mathematical formulas, to convey the essence of the fascinating works of energy to those not familiar with energetics.

The travel reports in Part I reveal the contours of the Staircase model. This is unavoidable because a scientist has processed the observations with an emerging model in his mind, a model that starts to act as a filter from the very moment of its conception. Paradoxically, this is actually a prerequisite for developing the model, but there is a danger that relevant observations will be filtered out too. Thus I may have missed some important observations. Maybe it is too ambitious to try to make a 'unified' model for the evolution of life. Nevertheless I have attempted this, because, as a human being, I am convinced that we must develop sustainably, and as a scientist, I believe that a good scientific model helps us to do this sensibly.

The evolutionary approach to history

For a long time, the history of human culture and that of biological species have been treated separately by rather distant scientific groups, adhering to different scientific philosophies, developing different scientific methods and speaking different scientific languages. History, according to the Encyclopædia Brittanica, is [1]:

> "The discipline that studies the chronological record of events (as affecting a nation or people), based on a critical examination of source materials and usually presenting an explanation of their causes."

Nonetheless, Johan Goudsblom [2] argues, such a cause-and-effect approach to historic developments yields no general insights about the past, or humankind as a whole; as a matter of fact it renders history *petite histoire*. Biologists, however, in following Charles Darwin, maintain and develop a systematic, evolutionary approach to (pre-)historic developments. In about one and a half centuries they have produced an impressive *Big History* of biological species from the origin of life more than three and a half billion years ago, until today.

In my view an evolutionary approach to historic developments, whether centred around living organisms or living organisations such as

PART I
THE FIRST FIVE ENERGY REVOLUTIONS

— a time journey through the history of life —

Hyperthermophiles are hard-nosed heat-lovers; they do not even survive moderate temperatures [22b]:

"Given it is unlikely that the whole ocean could stay at temperatures above 75°C for a sustained time ... the habitat of an early hyperthermophile may have been in a rock-hosted hydrothermal system."

This assumption also makes sense from a geological perspective. Even today, mid-ocean ridges can be found where *"extremely primitive cells ... cluster around vents of superheated water called black smokers"* [31]. These black smokers are not only sources of heat, but also of 'chemical food' consisting of sulphur-containing compounds; it appears that *"inconceivable numbers of sulphur bacteria thrive there, and form the base of a unique food chain"* [29b]. For consistency, the standard model implies that early cellular metabolisms were sulphur-based [22b].

Richard Fortey [29c] philosophises about early life and energy:

"The spark that is needed is energy, a supply of fuel to drive the whole motor of self-reproduction."

In the same vein, Nobel laureate Manfred Eigen says that energy supply was one of the primary demands on the evolution of early life [32a], which implied the development of a metabolic mechanism [32b]; no metabolism, no life. The earliest organisms extracted energy for the maintenance of their metabolism through fermentation of energy-rich inorganic compounds [32a]. They recruited phosphate compounds to channel the energy of life [24d]: adenosine triphosphate, in short ATP, emerged as *the* cellular distribution fuel of choice for biological work. The selection pressure must have been stringent, since only biochemical adaptations to the 'ATP fuelling technology'* survived all struggle for life. ATP became the principal cellular energy currency [33a] or the universal energy transmitter in biological systems [34a]. To this very day, any biological process, from cellular biosynthesis to muscle contraction or light generation by a glowworm, runs on ATP [35a].

---

*According to the Encyclopædia Britannica 2003, technology is: *"The application of scientific knowledge to the practical aims of human life or, as it is sometimes phrased, to the change and manipulation of the human environment."* In the proper sense of the word, Nature does not produce 'technology', but natural processes. Still I have opted for anthropomorphic word use, because I think that anthropomorphisms sometimes shed light on rather unexpected parallels between nature and human culture.

The Thermophilic Regime

Most likely sunlight played no principal role in originating life. Volcanic heat sources driving hydrothermal systems provided habitats for the early hyperthermophile organisms [22a]. Consequently, the prime energy source sparking off life was geothermal by nature [36a], that is primordial heat from the accretion events during the formation of planet Earth.* This notion is important for framing the 'Energy Time Scale' of the history of planet Earth proposed here. The first 'energy era' on the Energy Time Scale is the Thermoic. During the Thermoic, life originated through the emergence of thermophile organisms, or heat-lovers, the first ecologically dominant species on Earth which founded the first 'energy regime', the Thermophilic Regime.

PHOTOSYNTHESIS AND FREE OXYGEN

The origin of photosynthesis

The oldest actual fossils of microbial life date back some 3.6 to 3.5 billion years [22c], which is a few hundred thousand years after the first chemical signs of life occurred. These fossils consist of delicate chains of microbes that look exactly like blue-green bacteria (once misnamed as blue-green algae [7d]), or cyanobacteria, which are still in existence today [28b]. Blue-greens are so-called photosynthesisers; they collect sunlight and store the harvested solar energy. The evolutionary step that created them was perhaps a 'corporate merger' between two ancient bacteria, one of them delivering the sunlight-driven carbon-fixation system, while the other contributed the sunlight-driven oxygen-liberating centre. This latter mechanism probably evolved from a former sulphur bacterium that succeeded in metabolising dihydrogen oxide – or water – rather than dihydrogen sulphide [22c,27b,37a]. The digestion of water went with oxygen release, or oxygenesis.

As we have seen in the previous section, the standard model implies that the first living species were sulphur-processing organisms evolving in a hyperthermophilic environment. Subsequently, evolutionary innovations

---

*Currently it is thought that the first living cells were not only hyperthermophilic, but also chemolithotrophic, which means that both the energy source and the carbon source were inorganic. Raven and Skene [25c] talk about the *"chemolithotropic energization of the origin of life"*, but, as the authors elucidate, the inorganic chemical reactions were driven by the heat gradient built from a relatively hot hydrothermal vent and a relatively cold ancient ocean.

The Photo-Energy Revolution

in photosynthetic energy technology developed step by step. At first anoxygenic photosynthesis emerged, followed by oxygenic photosynthesis [22c,38].* Let us look more closely at the energy-conversion technology of photosynthesis, because that helps, I believe, in grasping the true nature of the energy revolutions that follow later.

Sunlight-driven energy technology

Solar energy travels through the cosmos in the form of solar radiation, which cannot be stored as such in, for instance, a 'sunlight container'. To store solar energy, it needs to be converted into an energy form that can be contained. The blue-greens invented a chemical means of doing just that. Let me clarify the essence of their great invention through a comparison with a so-called wind-water machine. The blue-greens managed to convert a flow of energy into an energy resource. In our wind-water model, the flow of energy is not sunlight but wind. Wind cannot be stored as such, in a 'wind container'; so the wind energy needs to be converted into a form that can be contained. The wind-water machine does this by using a windmill to harvest the wind energy. The mill drives a rotor, which in turn drives a water pump. The pump lifts water to an elevated storage reservoir. When the valve at the bottom of the storage reservoir is opened a flow of water develops which can drive a water wheel for the delivery of mechanical work. The wind-water machine thus converts wind energy, which cannot be stored, into contained 'water energy'. In effect, it converts kinetic energy into potential energy.

Basically the blue-greens did the same trick with solar and chemical energy as the wind-water machine does with wind and water energy. The typical colour of the blue-greens comes from light-sensitive green pigments, named chlorophylls [33b]. Chlorophylls are actually 'molecular solar panels' that harvest solar energy. The panels drive a molecular electron pump that lifts electrons to an elevated electron level shaped by

---

*According to the standard model, the last common ancestor(s) existed 4.2 to 3.8 billion years ago. While there is probable evidence for oxygenic photosynthesis about 3.5 billion years ago [22c], oxygenic photosynthesis was definitely occurring in cyanobacteria about 2.7 billion years ago [37b]. The lower time range seems to leave a short period for bacterial evolution to develop, but even 300 million years *"is an enormous length of time"*. Yet, we must always keep in mind that *"the model cannot be proven correct"*. It can be falsified, albeit that *"to date, the model has passed the tests"*. Moreover, *"it seems improbable that the sophisticated biochemistry needed for photosynthesis should spring out of nothing"* [23].

a chemical bond. Just like water, electrons can fall from a higher to a lower level, and just like falling water they then deliver work – chemical work in this case. A photosynthesiser converts solar energy that cannot be stored into contained chemical energy. Early blue-greens 'invented' a revolutionary, integrated solution for energy harvesting and storage that we call photosynthesis.

Photosynthesis technology embodies a sunlight-driven electron pump. By harvesting solar energy, blue-greens managed to strip electrons from water molecules which yields hydrogen nuclei, or protons, and molecular oxygen. The released hydrogen nuclei acquired an essential role in the blue-greens' biosynthetic activities, but oxygen was simply released as a waste gas. The impact of this emission can only be underestimated, because in those days the environment contained no oxygen. But before going more deeply into the fate of oxygen, we will follow the stripped electrons, as they hold the key to photosynthetic *energy storage*.

A wind-water machine, as discussed above, obviously is built before operation. Every part of it, including the water storage reservoir, has been prefabricated. However, photosynthesisers *synthesise* their electron storage reservoirs during operation; hence, the name photo*synthesis*. They pluck the necessary building materials from the environment on the run. To store the elevated electrons, blue-greens merge carbon dioxide and hydrogen nuclei in the making of sugars [33b,34c]. Overall, photosynthesis is a sunlight-driven chemical conversion of carbon dioxide and water into carbohydrates and oxygen. Photosynthetic carbohydrates form what you may call a 'photosynthetic crude', or a 'sugar crude'.

A sugar crude is completely combustible and therefore a fuel. But microbes need matter to grow and reproduce. To this end they, by-and-large, exploit photosynthetically fixed sugar-carbon for in-cell bio-construction works. Noticeably, airborne carbon dioxide is the only carbon source blue-greens consume [34b]. In effect, blue-greens create organic tissue from inorganic matter, or life from inanimate nature. Since they do not take organic carbon, biologists name them self-feeders [35b] or autotrophs [34b]. Ecologists call blue-greens producers. In their jargon, a consumer is an organism that does not synthesise the organic nutrients it needs, but gets them by feeding on the tissues of producers, or of other consumers [39a].

The Phototrophic Regime

When incident sunlight instead of primordial heat began to energise life, a true energy revolution happened on Earth: the Photo-Energy

The Photo-Energy Revolution

Revolution;* 'photo' refers to light. The Earth entered into a new energy era: the Photoic. In the first period of the Photoic, the Photian, so-called phototrophs began to dominate the planet's ecology [22c]; 'troph' comes from the Greek 'trophē', which means nourishment; life began to 'feed' itself with light. The photosynthesisers established a new energy regime: the Phototrophic Regime.†

AND THE FACE OF THE EARTH CHANGED

Two grand carbon cycles

The early atmosphere in the Thermoic era contained mostly carbon dioxide, perhaps some nitrogen, but little molecular oxygen [17d]. Some four billion years ago, the carbon dioxide concentration in the Earth's atmosphere was comparable to that of Mars today, which is 95.3 percent [40]. But two concurrent grand carbon cycles, the geo-chemical and the bio-chemical, primed the large-scale burial of carbon dioxide as silicate–carbonate or bio-organic matter, respectively.‡ The winners of the evolutionary battle fought during the Photian Period, the blue-greens, became the chief drivers of the bio-chemical cycle. Their oxygenic photosynthesis technology most probably enabled an increase in global organic productivity by at least two to three orders of magnitude as compared to 'conventional' hyperthermophile pursuits [41].

---

*In 'Vital Dust', Christian de Duve called this grand shift in the history of life the Green Revolution, referring to the colour of the molecular solar system that actually absorbed the solar radiation, that is chlorophyll [24e].

†It would be more precise to speak about photoautotrophy rather than phototrophy. Phototrophy refers to energy input by light, and autotrophy means that only carbon dioxide is needed as carbon source [34b]. In photoautotrophy the two properties are merged, and this merger alone established ecological dominancy. But, as phototrophs dominate autotrophy, I have chosen to use the designations Photian Period (short for Photoautotrophian) and Phototrophic Regime (short for Photoautotrophic). Besides autotrophs, biological evolution yielded heterotrophs that need organic compounds as carbon source, like human beings do. By far the majority of the heterotrophs are chemotrophs, rather than phototrophs, because energy is provided by oxidisable compounds from the environment.

‡Almost all the carbon dioxide of the early atmosphere ended up in silicate-carbonate or bio-organic matter [40]. Today about 81 percent of the carbon is buried in sedimentary rocks and about 19 percent in deposits of fossilised organic remains [43a].

The self-inflicted crisis

The new energy source tapped in the Photo-Energy Revolution not only caused a 'population explosion', but also created revolutionary mobility opportunities for modern bacteria mastering the novel 'light technology'. They could leave the hot springs *"to break free from their volcanic prisons and colonize the Earth as a whole"* [31b] – and they did. Blue-greens held dominion for ages during the Photian Period. Biological activity took place largely in the oceans, while simultaneously entire continents were forming, colliding and rifting apart [42a]. Around 2.5 billion years ago, shallow and coastal inland seas developed on the relatively large continents. These waters provided just the right environment for the blue-greens to flourish [42b]. Algal mats could thrive and expand, leading to much higher rates of oxygen production. But the freed oxygen released into the atmosphere induced the next energy revolution in the evolution of life, because it was toxic to its emitters, the blue-green bacteria. In effect, the rulers of the Phototrophic Regime created their own greatest enemy!

– 2 –
The Oxo-Energy Revolution

THE ORIGIN OF AEROBIC RESPIRATION

The origin of breathable air

In his exquisite book 'Life: A Natural History of the First Four Billion Years of Life on Earth', Richard Fortey inimitably sketches the far origin of the air we inhale [29d]:

> "I should reiterate that time – 3,500 million years ago – in order to try to come to terms with its immensity. First, imagine each cell exhaling the merest puff of oxygen, such as would fill a balloon smaller than a pin head. Then imagine a world thick with such cells, billions of them, dividing and dividing again, and each time they divide another minute puff of oxygen is given to the air. Then this process continues through generations that can only be reckoned as numerous stars in the Universe. And for every generation a thousand billion tiny balloons of oxygen released ..."

And yet, despite this laborious puffing, it seems to have taken well over a billion years before the photosynthetic activities of the blue-greens started to change the Earth's atmosphere [37c]. For a long time, the oxygen molecules released were rapidly captured by chemical agents from rocks, volcanic gases and upwelling oceanic iron particles [44] – the ancient 'red bed' deposits of oxidised iron found all over the globe remind us [45a]. Only when this slow, but irreversible, oxidation process was complete could the oxygen liberated by the blue-greens actually started to contribute to breathable air [29d,44]:

Oxian
About
2.1 billion
years ago

Photian
About
3.8 billion
years ago

Thermian
About
4.2 billion
years ago

If, and whenever, they emerged during the Archaean (from 4 to 2.5 billion years ago), early aerobes must have had a hard time. Outside the 'oxygen oases' the Earth remained a hostile place for the oxygen-breathers ... until the global oxygen event occurred and set them free. Both Lenton and Knoll put reports about the 'oxygen revolution' as if it brought *"an 'oxygen holocaust' in which untold lineages of anaerobic micro-organisms perished"* into perspective [43r]. Knoll maintains that anaerobes – including photosynthesisers such as the blue-greens, adds Lenton [37g] – *"simply retreated beneath an oxygenated veneer of surface sediments and water"*; they *"retained their critical roles in ecosystem function, roles that they retain today"*.

However, for ages blue-greens had been the prime producers of 'carbohydrate crude'. And it was not only for themselves: blue-greens were the suppliers of first choice for most heterotrophic consumers [7k]:

> *"All photosynthetic organisms ... leak organic compounds from the prodigious bounty of their harvested light ... Motility, whether by swimming, crawling, gliding, or creeping, serves to ensure that the hungry being incapable of photosynthesis remains in the well-lit zones mandatory to the photosynthesizer."*

This tendency of 'heterotrophs' to stay close to photosynthesisers *"led to thriving communities of feeders in the sunlit surface zones of marine and fresh water"*. When, during the global oxygen event, the blue-greens were forced to dive away, these *"thriving communities of feeders"* must have been threatened, and presumably marginalised. Hence the 'oxygen event' was nothing less than a 'crisis', causing in the end the heroes of the Phototrophic Regime to be dethroned. The environmental changes triggered new selection pressures that stimulated successors to evolve. But, before continuing our evolutionary journey, I would like to make a short diversion and discuss the workings of aerobic respiration technology, because it is going to be a main feature of our unrolling map of evolving energy systems.

Oxygen-driven energy technology

In the sections above I have referred to aerobic respiration using the term 'advanced internal combustion technology', because it is *"the most rewarding means"* of utilising energy from 'food' [37g]. Storing solar energy in chemical bonds of carbohydrates, as the blue-greens did, is one thing, but using the stored energy in a balanced manner is quite another.

Carbohydrates release energy during combustion. Early life basically developed two carbohydrate combustion technologies: bacterial *fermentation* and *anaerobic respiration* [34d]. The first technology applies combustion agents generated within the cell, while the second one utilises oxidants from the environment other than oxygen. From an energetical point of view, however, molecular oxygen is by far the preferred combustion agent. Let me try to clarify this assertion through a comparison with the wind-water machine.

The waterwheel of the wind-water machine works best when it is at the lowest level in the water loop; then the flow of water delivers optimal waterpower. Placing the waterwheel at a higher level yields less waterpower. A similar phenomenon occurs on the molecular scale with electrons. Photosynthesis, as we have seen, proceeds through sunlight-driven electron pumping and generates carbohydrates from carbon dioxide and water with simultaneous oxygen release. The combustion of photosynthetic carbohydrates is a chemical reaction which, simply put, *"takes apart what photosynthesis brings together"* [49a]. Hence optimal combustion closes the loop by re-producing carbon dioxide and water, re-taking molecular oxygen from the environment and re-sourcing energy. But the energy re-source cannot be in the original form of solar radiation. Effectively the solar energy stored previously during photosynthesis in sugars is converted into biochemical energy: the 'electron-fall' from carbohydrate to oxygen during combustion is chemically coupled to ATP formation, just like the waterfall of the wind-water machine is coupled to a waterwheel. Otherwise, the whole procedure would be useless as ATP is the universal cell fuel – no ATP, no living cell.

Complete combustion with oxygen as oxidant would develop maximum chemical power. But for living organisms that is too much of a good thing, because it does not leave any carbon-containing building blocks for internal bio-construction works. The microbes living in ancient oxygen oases achieved a well-balanced, not entirely complete combustion technology: a small proportion of the photosynthetic carbohydrates was partially combusted to yield small building units for biosynthesis; the rest underwent complete combustion to release energy for fuelling the metabolic activities. Biologists call this technology 'aerobic respiration', or respiration with oxygen as combustion agent.

Through aerobic respiration, nature completed a sophisticated biological analogy of the simple artificial wind-water machine: the 'photon-electron machine'. The microbial equivalent of the water upstream is *oxygenic photosynthesis*, while the downstream process corresponds to *aerobic respiration*. To sense the magnitude of the aerobic respiration achievement in terms of energetics, we compare it with anaerobic combustion. Anaerobic combustion through fermentation yields approximately two cell fuel units, or ATP molecules, per glucose unit [34d]. This amounts to about 5 percent energy recovery based on crude carbohydrate [35d].* Aerobic respiration, on the other hand, yields 38 molecules of ATP per glucose unit [34e].† Based on photosynthetically stored energy this boils down to 80 percent efficiency [35d]. Thus, *aerobic* respiration extracts almost 20 times more energy from the same photosynthetic 'sugar crude' than *anaerobic* fermentation does.

MICROBIAL SYMBIOSIS

The origin of the eukaryotic cell design

We have seen how during the Photian Period bacterial evolution created two important metabolic innovations, each applying advanced energy conversion technologies. The first was *oxygenic photosynthesis*, technology for the conversion of solar energy into stored biochemical energy, and the second was *aerobic respiration*, technology for the conversion of stored biochemical energy into microbial work. During the Photian Period, oxygenic photosynthesis spread worldwide with the rising ecological dominance of the blue-greens. The blue-greens could emit oxygen for long periods of time without oxygenating the air, because the Earth took all the oxygen to become chemically oxidised. But at a certain juncture, the oxidation of the Earth was completed and the air was enriched with oxygen. That was a great event for aerobes, which needed oxygen to live, but a crisis for anaerobes as they are killed by oxygen. How sustainable could life have been in a world with blue-greens on the list of endangered species? For the blue-greens produced not only oxygen, but also food for respiring aerobes. Both the aerobic and the anaerobic energy technologies were at risk. However, this grand drama dissolved in time, enabled by the evolution of a whole new cell design.

To be able to report appropriately about the next trajectories of our evolutionary excursion I must introduce a few basic biological terms. Today's standard model of Darwin's great Tree of Life, a depiction of the genealogical relationships of all living organisms, has three principal branches: bacteria, archaea and eucarya [43f]. The basic unit from which all living things are composed is the biological cell defined as the *"the smallest structural unit of living matter that is able to function independently"*

---

*The ATP yield of *anaerobic respiration* depends on the particular oxidant used, but almost all oxidants are significantly less efficient than oxygen.

†This figure refers to the ATP obtained within the mitochondria and does not consider the energy requirement of ATP export into the cell [34e].

[46]. Biologists identify two fundamentally different cell organisations: the *prokaryotes* and the *eukaryotes*. Both bacteria and archaea appear to be prokaryotes; the Eucarya are eukaryotes [43c,f].

For over a billion years, the living world consisted of bacteria and archaea only, thus prokaryotes. But 2.7 billion years ago (or earlier!) the prokaryotic biosphere was augmented with a fundamentally new cell design: the eukaryotes. All visible organisms in our world – plants and animals – are eukaryotic by nature [7e]; indeed, the human body is also a living sculpture composed of eukaryotic cells. The prokaryotes did not go away; in fact the opposite is true: *"they maintain the foundation of all functioning ecosystems on this planet"* [43g]. For example, *"biology's thirst for nitrogen is quenched mainly by bacteria"* [43h].

The prokaryotic and eukaryotic cell designs differ in numerous aspects. The presence of internal membranes enclosing so-called organelles distinguishes eukaryotic cells from prokaryotic cells [47a]. An organelle has a specialised function [43i], the cell's defining one being the nucleus which *"contains the cell's genes. Inside the nucleus, long strands of DNA are wound tightly around tiny proteinaceous beads to form linear chromosomes."* Prokaryotes lack organelles, and thus lack a nucleus. They have a design based on one compartment that is largely filled with a liquid called cytoplasm. Their DNA is usually a single stranded circular molecule. But the design may not be the most important difference, as Andrew Knoll explains [43i]:

"Perhaps the most important difference between eukaryotes and other cells concern the way in which the cell's contents are stabilized. Archaeans and bacteria enclose their cytoplasm in a rigid wall. In contrast, eukaryotes evolved an internal scaffolding called the cytoskeleton, and that, as Robert Frost once wrote, has made all the difference. Built from tiny filaments of actin and other proteins, the cytoskeleton is a remarkable dynamic structure, continually able to form and re-form in ways that change the cell's shape. Many of us remember film clips of amoebas, viewed on slow days in high school biology class. The graceful undulations of the amoeba, its pseudopodia extending to capture prey, reflect the coordinated action of a dynamic cytoskeleton and a flexible membrane system. This coordination is key to the evolutionary success of eukaryotes precisely because it enabled these cells to engulf particles."

A putative descendant of the ancient eukaryotes is *Giardia* [24g], a giant – in the 'eyes' of a blue-green – single-celled organism. Its volume exceeds that of an average prokaryote by more than 10,000 times. It is not encased within a solid membrane, like bacteria, but coated only by some fuzzy material without rigidity. *Giardia* swims well, being equipped with four pairs of long flagella. Most remarkably, *Giardia* swallows up whole bacteria including genome and metabolic features such as energy and food-processing technology.

Margulis and Sagan view 'eukaryosis' as *"the transition from bacteria to our kind of (that is, eukaryotic) cells"* [7e]. They argue that [7l]:

> *"Evolutionary novelty of the nucleated cell is best comprehended as specific historical products of partnerships and symbioses, bacterial fusions of DNA whose products (proteins, RNA's, lipids) interact to generate emergent structures."*

Simply put, *"eukaryotes acquire and integrate entire complete genomes to form 'individuals'"* [7i] from microbial 'symbionts', the microbial partners in symbiotic alliances.*,†

The microbial energy majors

According to Andrew Knoll, *"in a 1967 paper that I later learned was rejected fifteen times before finding a home"*, Lynn Margulis reinvented the endosymbiotic hypothesis for eukaryotic cell origins. She then proposed that the mitochondria and chloroplasts, organelles of a green plant cell, *"were descended from free-living bacteria"* – an oxygen-respiring bacterium and a cyanobacterium, the blue-green, respectively [7i] – which today biologists accept as fact [43j].

The first aerobic eukaryotes must have carried enzymatic tools to detoxify reactive oxygen products, otherwise they would not have been capable of running an oxygen-based metabolism. They could thus defend a symbiont like a blue-green against toxic oxygen derivatives. Through symbiosis the threatened oxygenic photosynthesis technology was not only saved and secured, but also prepared for further evolutionary development. In the photosynthesising eukaryotic cell the chloroplasts extract solar energy, store it in sugars and supply the cellular host; the host, in turn, delivers the sugar to the mitochondria that in turn copiously supply the host with cell-fuel (or ATP). It happened to be a revolutionary good deal. Richard Fortey, expressive as always [29f]:

> *"The [Eukaryotic] cell can be compared with an engine run by chemical fuels, but powered by motors that were capable of running on their own. Life hoisted itself up by its own bootstraps."*

---

*'Endosymbiosis' is also used to indicate the typical cellular partnerships with one cell nested inside the other [43j].

†Margulis and Sagan suggest that in the start-up phase so-called 'bacteria consortia' – *"bacteria living with and inside other bacteria"* – *"almost certainly"* induced the origin of the first eukarytic cell [7g,h].

The Oxo-Energy Revolution

The origin of the integrated solution exploiting symbiotically oxygenic photosynthesis and aerobic respiration marks a new energy revolution: the Oxo-Energy Revolution; 'oxo' refers to oxygen. The Oxo-Energy Revolution initiated the establishment of a new energy regime, the Aerobic Regime, which reigned during the Oxian Period. Complex eukaryotes, equipped with mitochondria and additional chloroplasts in the case of a photosynthesiser, became ecologically dominant. The historic alliance between 'light-eaters' and 'oxygen-breathers' not only rescued the blue-green's light-harvesting technology, it also increased the energy efficiency of photosynthetic, cellular metabolism by an order of magnitude. Without this quantum leap in solar energy utilisation, bio-processes would have been too sluggish to sustain macroscopic life [17e]. Hence the Oxo-Energy Revolution was a true solar energy revolution enabling nothing less than the emergence of what we today perceive as Nature.

The Aerobic Regime

The oldest biomarkers – well-preserved molecular fossils – assigned to eukaryotic cells are 2.7 billion years old [43k]. Yet the earliest appearance of eukaryotic photosynthesisers equipped with aerobic respiration technology was made around 2.1 billion years ago [37g].* This coincides with the global oxygen event that marks the beginning of the Oxian Period on the Energy Time Scale.† But the Oxo-Energy Revolution needed time to mature.

The palaeontologist Andrew Knoll observes [43l]:

"By 1.5 billion years ago, then, the cyanobacterial revolution may have been complete, but a second revolution – the rise of eukaryotes to ecological prominence – was yet to come."‡

---

*Lenton adds that recent genomic sequence analyses suggest a later origin of aerobic eukaryotes, namely 1.8 billion years ago [37g].

†After this rise in oxygen, the paleo-record reveals an interval of relative stability lasting more than one billion years [37c]. To explain this, Nisbet and Sleep imagined a balance between the carbon dioxide assimilation and the oxygenesis of photosynthesis with the carbon dioxide emission and oxygen use of aerobic respiration [23]; this agrees with the origin of eukaryotic photosynthesis coinciding approximately with the global oxygen event from 2.2 to 2.0 billion years ago.

‡A word about the timing of the 'revolution'. Knoll dates the 'Eukaryotical Revolution' as being at the threshold *"of the rise eukaryotes to ecological prominence"*, which makes perfect sense, but, for reasons of consistency, I place the milestone at the origination of the next ecologically dominant agent – in this case the photosynthetic eukaryotes. This is akin to how historians mark the beginnings of the Agricultural Revolution and the Industrial Revolution.

Knoll identifies a puzzling question [43m]:

> *"If the early course of eukaryotic evolution was set so early in life's history, why should the domain (our domain) have remained subservient to prokaryotes for a billion and a half years before spreading throughout the oceans?"*

Taking a look at the environment he concludes that the first photosynthetic eukaryotes, or algae, *"compete poorly against blue-greens for the scarce nitrogen compounds that would have been present in mid-Proterozoic seawater"* [43n]. Algae cannot fix nitrogen like the blue-greens. To get nitrogen, algae must digest nitrate, but that was a scarce nutrient in the mid-Proterozoic ocean. Only when this nitrogen limitation began to weaken, about 1.2 billion years ago, could seaweeds and plankton start to diversify. Hence the full evolutionary development of eukaryotic organisms dates from less than 1.2 billion years ago, when the *"the interaction between genetic possibility and environmental opportunity"* smoothed further progress of speciation [43n].

Biological diversification

One group of the newly evolved complex eukaryotes, those equipped with photosynthetic technology, developed into the green plants that surround us today [24f]:

> *"When you admire the greenery of a tropical forest, you are seeing blue-greens propelled to unprecedented ecological success by hitchhiking in a protozoan."* [43o]

Eukaryotes that did not fuse with blue-greens but 'only' respired aerobically evolved to moulds and animals, and finally to humans [24f]. The eukaryotic cell design enabled a huge enrichment in cellular diversity accelerated by sexual reproduction [29e]. Exchange of genetic material led to the enhancement of natural variation within a species, which, in turn, induced greater adaptability [29f]. Moreover, eukaryotic organisms developed intracellular regulatory systems by *"molecular signals that pass from cell to cell, switching specific genes on or off as they go"* [43p]. The corresponding evolutionary advancement of cellular 'information communication technology' co-enabled the evolution from single-celled to multicellular organisms so co-fueling the expansion of eukaryotic diversity; *"ninety-five percent of extant eukaryotic species are multicellular"* [43p].

Around 1.8 billion years ago the first multicellular fossils appeared *"in the form of mysterious coils measuring about 2 cm long. The oldest recognisable multicellular organisms are red algae dating back to 1.2 billion years."* [28e]

The Oxo-Energy Revolution

The trend continued towards increasing morphological diversity [17f]. As discussed above eukaryotes swallow particles, including cells. In effect, they *"introduced grazing and predation into microbial ecosystems"*. Seven hundred and fifty million years ago, building on this capacity, *"eukaryotes had begun to construct the complex food webs that today form a crown – intricate and unnecessary – atop ecosystems fundamentally maintained by prokaryotic metabolism ... Earth was becoming a eukaryotic planet."* [43q]

Our own multicellular lineage – the animals – passed down fossils about 575 million years old, although, says Carl Zimmer, 'animal' *"is a generous description"* for a collection of coin-like, mostly immobile creatures [28e]. About 40 million years later species diversification accelerated during the so-called Cambrian Explosion, which took place in only 10 million years – *completely under water* [28e]. According to David Jablonski [48]:

"Regardless of exactly when the major metazoan lineages actually diverged, the Cambrian explosion represents a uniquely rich and temporally discrete episode of morphological evolution. Almost all of the skeletonized metazoan phyla appear within an interval of perhaps 10 million years, and the accompanying diversification of microplankton, of forms having agglutinated skeletons (i.e. pieced-together sand grains or skeletal debris), and of behavioral traces all suggest that this is a real evolutionary event and not simply a change in the preservation potential of an already diverse biota."

It is far from clear what precisely caused the Cambrian explosion to occur, but a substantially increased oxygen availability as compared to the values reached during the long stasis after the global oxygen event *"may have been a critical prerequisite for the evolution of large, complex metazoans"* [48]. The continents were bare in those days, except for bacterial crusts. But, not too long after the Cambrian Explosion, around 500 million years ago, the first plants spread onshore [28e]:

"The first land plants probably looked like today's moss and liverworts, forming a low, soggy carpet along the banks of rivers and coastlines. By 450 million years ago centipedes and other invertebrates were beginning to explore this new ecosystem. New species of plants evolved that could hold themselves upright, and by 360 million years ago trees were growing 60 feet high. Out of the coastal swamps would sometimes slither our ancestors – the first vertebrates that could walk on land."

The incredible story of the *biological evolution* of species suggests that 'diversification' was the order of the day for hundreds of millions of years – which is true, but only half true. An *evolutionary energetics* approach balances that view. According to Margulis and Sagan [7m]:

> *"The entire animal kingdom employs essentially one mode of metabolism: the use of oxygen to respire organic food molecules. This is called heterotrophy. Plants employ two: heterotrophy via oxygen just as in animals, and carbon-dioxide-fixing, oxygen-producing photosynthesis by use of the sun's radiation. This is called photoautotrophy. Bacteria, in addition to heterotrophy and photoautotrophy, have at least twenty metabolic modes that are not possessed (except via symbiotic acquisition) by either animal or plant."*

The advanced eukaryotes almost entirely threw their lot in with an integrated solution of 'only' two ancient bacterial energy technologies: oxygenix photosynthesis by chloroplasts and aerobic respiration by mitochondria [7j]. The first trees did, so did the dinosaurs, and so do we. In 'Symbiotic Planet: A New Look at Evolution' Lynn Margulis writes [51a]:

> *"Wherever we humans go, the mitochondria go too, since they are inside us, powering all our metabolism: that of our muscles, our digestion, and our thoughtful brains."*

Ecological growth

Life created a global oxygen pump, a photochemical oxygen pump shaped by co-evolving biological and environmental processes, driven by the sun. Even today practically all the oxygen present in the atmosphere comes from photosynthetic organisms [52a]. Together, eukaryotic photosynthesis and respiration cycle carbon through the biosphere, *"sustaining life and maintaining the environment through time"* [43l]. But a robust energy economy does not imply ecological stasis, or no-growth. We have seen how precisely the opposite happened in the history of Nature: just because of new energy technology, Nature's economy grew phenomenally during the Aerobic Regime. Innumerable food chains (says the ecologist), or energy chains (prefers the energeticist), formed complex ecosystems. Interestingly, innovations in mobility technology appeared to be crucially instrumental. Organisms with skeletons originated in rapid succession during the Cambrian Explosion [42b,c,31b]. That not only enabled animals to maintain significantly larger bodies, it facilitated the development of advanced locomotive functions. Most of the energy invested in locomotion was used in the search for food [53a]. A diversity of predatory-prey food chains emerged: bacteria never passed one level – predator invaded prey, and that was it – but aerobes created many complex, multilevel food chains in which smaller species were preyed upon by larger species, which subsequently were eaten by their even larger brethren [42d].

At the beginning of a food chain there is always a food producer, such as photosynthesising green plants. Then come consumers such as herbivores, carnivores, and omnivores. Eventually, all consumers depend on energy-rich green substrates for living, reproduction, and locomotion, or biosynthetical and biomechanical activities. All species, including us, are made out of water, carbon dioxide, and some minerals gathered and prepared for consumption by sunlight-harvesting green plants.*

AND THE FACE OF THE EARTH CHANGED

The changing landscape

During the Aerobic Regime plants and animals changed the face of the Earth quite radically. They set foot on the continents and completely remodelled the original, desolate scene. Life made land 'productive' by building biological factories [31c]. In the words of Richard Fortey [29f]:

> "Life made the surface of the Earth what it is, even while it was Earth's tenant."

After about one and a half billion years oxygenic aerobes had transformed the atmosphere from oxygen-poor to oxygen-enriched. With a rising oxygen level, the ozone layer arose [45a]. Ozone shields the Earth's surface from ultraviolet solar radiation. That promoted the emergence of life forms at the continents; otherwise, species would have had to develop UV protection mechanisms themselves [17f]. Of course other events, for

---

*Many of us were taught at school that green plants fix carbon dioxide and give off oxygen, while animals inhale oxygen and exhale carbon dioxide. This is true. But plants inhale oxygen and exhale carbon dioxide too, because they are aerobes. We have seen that eukaryotic cells are symbiotic alliances. The cells of the plants carry organelles for solar energy harvesting and storage, the chloroplasts, and organelles for energy recovery, the mitochondria. As usual in alliances, the organelles work independently together. The plant's aerobic respiration activity depends on species and growth stage: it ranges from 20 percent in young, rapidly growing plants to over 90 percent in mature forests [33c]. In the latter ecosystems plants need almost all the harvested solar energy and fixed carbon to support their own metabolism. In terms of net photosynthetic activity, the highest recorded field values are 4 to 5 percent during short periods with optimum conditions. The weighted global mean, pulled downwards by the mature forests, is just around two-tenths of one percent [33c].

instance astronomical or geological by nature, influenced the evolutionary course of the Earth's ecosphere. Evolution is the outcome of co-evolutionary processes that develop in separate niches but also interact.* During the Aerobic Regime several dramatic climate changes passed. Part of the Earth's climate record, which shows nine ice ages, can be explained by astronomical cycles, such as alterations in the Earth's orbit around the sun on a 100,000-year timescale affecting the amount of solar radiation that reaches our planet [31b]. Further, meteorite impacts, continental plate tectonics, and periods of intense volcanic activity profoundly affected the evolution of life. They played their role (in addition to other triggers changing the oceans and the atmosphere) in catastrophic waves of extinctions, with five standing out in particular [28g]:

"They have ripped through the fabric of life, destroying as many as 90 percent of all species on Earth in a geological instant ... And once mass extinctions strike, it takes millions of years for life to recover its former diversity. In the wake of mass extinctions life can change for good. They can wipe out old dominant forms and let new ones change for good. In fact, we may owe our own success to such shifts of fortune ... Most species live between 1 and 10 million years, and new species come into existence with roughly the same rate as older species disappear."

The two dominant energy technologies, oxygenic photosynthesis and aerobic respiration carried by complex eukaryotes, survived all mass extinctions.

Competition and adaptation

Evolution has been shaped by a fundamental tension between two key agents of natural selection: *competition* for food and *adaptation* to changes in the physical environment. Perhaps competition dominates in the long periods between severe environmental changes [31d]. Organisms then evolve to ever-larger sizes. But rapid environmental change breaks the vicious cycle of competition, and forces species to adapt. And the better the species' adaptations are, the higher its chances of survival. Two billion years ago, the then dominant species, photosynthesising oxygenic

---

*The concept of co-evolution, like the concept of evolution itself, is applicable to macro-developments such as cosmological, biological and cultural evolution as well as, for example, to the development of a distinct species. The co-evolutionary way in which the evolution of one species co-shapes the evolution of another is one of the most powerful forces shaping life [28f].

The Oxo-Energy Revolution

blue-greens, caused an environmental catastrophe of immense size: the global oxygen crisis. A remarkable adaptive innovation conquered ecological dominance thereafter. Through 'strategic' symbiosis energy-efficient oxygen-breathing bacteria and light-harvesting blue-greens – which were suffocating in their own waste – participated in new cell designs. Enabled by the revolutionary synergism between the energy technologies involved, these microbial mergers, or complex eukaryotic cells, could grow impressive plants and animals. Photosynthetic and non-photosynthetic eukaryotes created a world that was natural ... until genus *Homo* initiated the next energy revolution.

– 3 –
The Pyro-Energy Revolution

THE ORIGIN OF HOMINIDS

It needed a crisis

Around 250 million years ago, the dinosaurs emerged. These colossal aerobes evolved into the dominant land animals of their time, which lasted until 65 million years ago. Then, suddenly, most of their branches became extinct [28h,i]. Probably a giant asteroid – with a diameter of about 10 km and a speed somewhere around 27 km a second – disturbed life on Earth quite energetically [28j,k,31e]. Carl Zimmer put it into words as follows [28k]:

> "The impact sprayed rock and asteroid 100 kilometres into the sky, above the stratosphere. An earthquake 1,000 times more powerful than anything in recorded history made the entire planet shiver ... The world burned; smoke hid the sun. Plants and phytoplankton died in the prolonged darkness, and the ecosystems that were built on them collapsed ... When the skies cleared after the impact ... the giants were gone."

A coincidental meeting of planet Earth with another celestial body changed the course of biological evolution. One species' breath is another species' death. According to palaeontologist Peter Ward [28k]:

> "It was the removal of the dinosaurs through mass extinction that allowed so many lineages of mammals to come about through the evolutionary process ... There would not be humans here but for that mass extinction."

Talk about co-evolution! Mammals lived alongside the mighty dinosaurs for just under 200 million years. But it took a major asteroid intervention before they could

---

Pyrian
About
0.5 million
years ago

Oxian
About
2.1 billion
years ago

Photian
About
3.8 billion
years ago

Thermian
About
4.2 billion
years ago

make the most of their evolutionary potency [28h,i]. Of course, they suffered along with the other life forms. Many marine animals became extinct, as did all land animals weighing more than about 25 kg [31e]; perhaps two-thirds of all mammalian species disappeared. But the ones that survived inherited the Earth, and within 15 million years they had evolved into the forbears of the mammals alive today. The primates were among these [28k], making their appearance as athletic tree-dwellers.

About 25 million years ago, the first ape-like primates showed up in the tropical rain forests of East Africa [31f,54a]. But after more than 15 million years [31f,42e], global cooling induced a thinning of the forests and gradually destroyed their traditional habitats [55a]. Grasslands, on the contrary, expanded [31f]. Entrepreneurial tree-dwellers grasped the opportunity to exploit the food-rich pastures, and diverged from the chimpanzees [31f,54a,55b]. The land scouts evolved into the human line, or the family *Hominidae* [55c], defined by the habit of walking upright, on two legs [54b]. With food patches more widely dispersed, selection pressure favoured more efficient locomotion [55a]. And, logically, the two-footed primates moved away from their original ape-like diet [55b].

The origin of human culture

The initial advantages of walking upright may always remain a matter for speculation, but they must have been significant. A progressively cooler and drier environment again necessitated other survival strategies. Ape-like bipeds adapted to a less regular food supply by developing strong cheek-teeth. This enabled them to cope with hard fruit and nuts [31f,55b]. But, as Richard Klein in 'The Dawn of Human Culture: A Bold New Theory on What Sparked the *"Big Bang"* of Human Consciousness' describes [54c]:

> *"Around 2.5 million years ago, some scrawny bipedal creature made a revolutionary discovery that greatly increased its chances for survival. It lived in woodlands or savannas where predators, accidents, disease, or starvation killed antelopes, zebras, wild pigs, and other large mammals. Carnivores and scavengers did not claim all the available flesh or marrow, and therein lay an opportunity. What our spindly biped found was that if it struck one stone against another in just the right way, it could knock off thin, sharp-edged flakes that could pierce the hide of a zebra or gazelle. It could use the same flakes to slice through the tendons that bind muscle to bone. In effect, it had found a way to substitute stone flakes for the long slicing teeth that cats and other carnivores employ to strip meat from a carcass ... In extending their*

anatomy with tools so that they could behave more like carnivores, they set in train a co-evolutionary interaction between brain and behaviour."

Archaeologists group similar stone tool assemblages within a so-called 'Industry', or 'Culture'. That way all assemblages from between 2.5 and 1.6 million years ago, which appear to closely resemble each other, are assigned to the so-called Oldowan Industry [54i]:

"By later human standards, Oldowan stone-working technology was remarkably crude, and an observer might reasonably ask if it exceeded the capability of a chimpanzee. The answer is probably yes ... the evidence so far suggests that even an especially intelligent and responsive ape cannot grasp the mechanics of stone making."

With the appearance of Oldowan tools the evolutionary lines of bipeds parted: one developed into the robust, smaller-brained and larger-toothed australopiths and the other into the relatively slender, larger-brained and smaller-toothed genus *Homo* [54d,e,55b].* The use of stone cutlery made a significant difference: it enabled early humans to add animal flesh and marrow to a mainly vegetarian diet [54b], and possibly this stimulated the metabolic activity necessary to support a larger brain [31f]. Our ancient forefathers also developed a muscled body that allowed for greater activity, including the capacity to run [55b].

Around 1.7 million years ago, *Homo ergaster* – or 'working man' – entered the evolutionary arena as maybe the first *"true human"* [54b,g]. His arrival coincided with technological innovation: Oldowan Industry made way for the so-called Acheulean Industry, characterised by sophisticated hand axes [54h,i]. Both the tools and their makers persisted largely unchanged for about one million years. Then, about half a million years ago *"the glacial monotomy of life began to break up"* [28l]. The hand axe advanced, regional styles appeared, and humans learned how to make spears. This discontinuity concurred with a relatively abrupt increase in human brain size [54f,j], and the swan song of *Homo ergaster*; *Homo heidelbergensis* had arrived [54g]. He appeared in Africa some 600,000 to 500,000 years ago and successfully colonised Europe over the next 100,000 years [54g,k], perhaps as a result of his improved late Acheulean stone-tool innovation [54f].

Homo heidelbergensis had a large brain: about 90 percent of the modern average; in contrast his predecessor, *Homo ergaster*, had achieved only about 65 percent [54f] of the modern average. Anyway, a modern human

---

*In his recent book 'The Dawn of Human Culture', Richard Klein says: *"The scientific evidence for human evolution has been accumulating for more than 150 years, and much has been added in just the past decade. The sum now allows a broad outline that is likely to stand the test of time."* [54f]

brain is not only large, but also exceptionally heavy: the brain-to-body ratio is roughly six times larger than that of other mammals, and about three times larger if the survey is confined to monkeys and apes. It looks like the brain evolved more rapidly than any other organ in the history of the vertebrates [54f] (i.e. a species possessing a spinal column); early vertebrates appeared about 450 million years ago [31b] in the game of natural selection. We always highlight the benefits of our large brain, but there is a price to pay: energetic costs; in modern humans the brain consumes roughly 20 percent of the body's metabolic resources [54f], and brain tissue demands 22 times more energy than an equivalent piece of muscle at rest [28m].

The quest for the human advantage

Despite his big brain, *Homo heidelbergensis* was an ineffective hunter, less effective as a predator or scavenger than, for instance, the hyena [54l]. In two million years of evolutionary development, 'Man the Toolmaker' [39b, 55d] failed to establish ecological dominance over other animals. Something beyond tool making must have happened during the last half million years that made humans suddenly successful in colonising the planet. Many scholars rack their brains over the origin and nature of the human advantage. Roger Lewin [55d] proposed an intriguing hypothesis:

> *"Perhaps ... the evolution of language, not the development of technology, provided the evolutionary foundation upon which natural selection built the bigger brain?"*

But, argues Richard Klein [54m]:

> *"No topic is more intriguing and more difficult to address concretely than the evolution of language, but ... [it] is almost a kind of sixth sense, since it allows people to supplement their five primary senses with information drawn from the primary senses of others. Seen in this light, language becomes a kind of "knowledge sense" that promotes the construction of extraordinarily complex mental models, and language alone may have provided sufficient benefit to override the costs of brain expansion."*

Without having fossilised 'word artefacts' in front of him, the neuroscientist Terrence Deacon thoughtfully elaborates on the 'Man the Talker' theme. In his book 'The Symbolic Species: The Co-evolution of Language and the Brain', he explains how important it is to discriminate between communicating and talking [56a]:

> *"Although other animals communicate with one another, at least within the same species, this communication resembles language only in a very*

superficial way – for example, using sounds – but none that I know of has the equivalents of such things as words, much less nouns, verbs, and sentences. Not even simple ones."

And that is amazing, he says [56a], because:

"Most mammals aren't stupid. Many are capable of quite remarkable learning. Yet they don't communicate with simple languages, nor do they show much of a capacity to learn them."

Deacon claims that the defining attribute of human beings is an unparalleled cognitive ability, as our species designation 'sapiens' suggests [56b]:

"We think differently from all other creatures on earth, and we can share those thoughts with one another in ways that no other species even approaches. In comparison, the rest of our biology is almost identical. Hundreds of millions of years of evolution have produced hundreds of thousands of species with brains, and tens of thousands with complex behavioural, perceptual, and learning abilities. Only one of these has ever wondered about his place in the world, because only one evolved the ability to do so.

Though we share the same earth with millions of living creatures, we also live in a world that no other species has access to. We inhabit a world full of abstractions, impossibilities, and paradoxes ... We tell stories about our real experiences and invent stories about imagined ones, and we even make use of these stories to organize our lives. In a real sense, we live our lives in this shared virtual world. And slowly, over the millennia, we have come to realize that no other species on earth seems able to follow us into this miraculous place.

... The doorway into this virtual world was opened to us alone by the evolution of language, because language is not merely a mode of communication, it is also the outward expression of an unusual mode of thought – symbolic representation. Without symbolization the entire virtual world ... is out of reach: inconceivable. My extravagant claim to know what other species cannot know rests on evidence that symbolic thought does not come innately built in, but develops by internalising the symbolic process that underlies language. So species that have not acquired the ability to communicate symbolically cannot have acquired the ability to think this way either.

... [Language] entirely shapes our thinking and the ways we know the physical world."

But although *Homo heidelbergensis* had a big brain, all the signs are that he was not at all an advanced symbolic thinker. The point is: brain and mind are not the same. The brain is an *organ*, and the mind is its *capacity* to feel, perceive, think, will, and especially to reason – the mind is 'brainpower' [56b]:

> *"It is not just the origins of our biological species that we seek to explain, but the origin of our novel form of mind. Biologically, we are just another ape. Mentally, we are a new phylum of organisms."*

Deacon implies that the quest for what made humans different from animals boils down to a search for the origin of the human mind, or rather the symbolically thinking human. He infers that thinking symbolically and communicating linguistically are different sides of the same coin. Both require people to learn and perform some remarkably complicated skills, such as the production of speech and the analysis of speech sounds [56c]. But [56d]:

> *"When we strip away the complexity, only one significant difference between language and nonlanguage communication remains: the common, everyday miracle of word meaning and reference ... Just the simple problem of figuring out how combinations of words refer to things."*

Deacon argues that the evolution of symbolic reference initiated the co-evolution of language and brain, the trigger being a single symbolic innovation that was initially extremely difficult to acquire. This exerted a selection pressure on the brain making the symbolic threshold easier to cross. More brainpower in turn enabled a greater capacity to symbolise, and thus to speak, and think. Simple languages and brains so co-evolved into complex ones [56d]. But what kind of environment caused, nurtured, and selected the act of *symbolic referencing*, or speaking language? A symbol system, argues Deacon, demands *"significant external social support in order to be learned"* [56e]. Thus [56f]:

> *"Language is a social phenomenon ... [and] ... The relationship between language and people is symbiotic."*

In his search for *the* social differentiator, Deacon assumes that early humans were dual-parenting species. They tended to form cooperative pairs with strong, exclusive attachments to one another [56g]. The clue to the forces that shaped their unique social communication must thus lie in the way they negotiated reproductive activities. Many species seem to *"do it chemically"*: one partner produces chemicals in order to elicit a sexual response in the other, the fascinating chemicals being sex pheromones. But the chemical approach did not suit our forefathers, Deacon reasons [56h]:

> *"Unfortunately, as primates with diminutive smell organs, we humans are not well equipped for smell-governed behavior. It should not be surprising,*

The Pyro-Energy Revolution

> *then, to discover that these intrinsically unstable reproductive arrangements in human societies are stabilized by a unique form of social communication, which can be as powerful and reliable as a social hormone."*

In addition [56i]:

> *"Males must hunt cooperatively; females cannot hunt because of their ongoing reproductive burdens; and yet hunted meat must get to those females least able to gain access to it directly (those with young), if it is to be a critical subsistence food. It must come from males … [who] … must maintain constant pair-bonding relationships."*

It looks like the hunting-for-meat survival strategy promoted a distinct social structure implying a symbolisational solution [56i]; symbolic reference is intrinsically social [56j], and thus speaking and thinking are too.*

Deacon makes it plausible that the evolution of symbolic reference shaped the co-evolution of language and brain through a process of socialisation. Nevertheless, the co-evolution of social evolution and biological evolution may not be overlooked. Having to say something is one thing, but being equipped with a suitable voice box is another. One needs a rather peculiar sort of anatomy in order to speak, an anatomy that developed with humans [28n].

Deacon infers that the human advantage is the capacity to be a more socialised animal than other social animals are. And he is not alone in thinking this [28o]:

> *"Compared to animals, we are a supremely social species. We live in a global network of nations, alliances, tribes, clubs, friendships, corporations, leagues, unions, and secret societies,"*

to quote from 'Evolution: The Triumph of an Idea'. With regard to social evolution, it further reads [28o]:

> *"Yet as difficult as it is to glimpse the social evolution of humans, scientists suspect that it was a crucial factor in the rise of our species – perhaps the*

---

*The brain is an organ carrying the elusive mind. The mind holds the even more indescribable consciousness, a thought-provoking subject of long standing for philosophers, though the concept of symbolic referencing helps to make the miraculous consciousness describable. According to Deacon [56k]: *"Consciousness of self … implicitly includes consciousness of other selves, and other consciousnesses can only be represented through the virtual reference created by symbols … It is a final irony that it is the virtual, not actual, reference that symbols provide, which gives rise to the experience of self. This most undeniably real experience is a virtual reality."*

crucial factor. Our chimplike ancestors had chimplike social lives, but five million years ago they branched away from other apes and began to explore a new ecological niche on the savannas of East Africa where their social lives became far more complex. Much of what makes humans special – our big brain, our intelligence, even our gift of language and our ability to use tools – may have evolved as a result. At the same time, the competition for mates and the struggle for reproductive success among these hominid ancestors of ours may have left their mark on our psychology, shaping our capacity for love, jealously, and all other emotions."

Robin Dunbar suggests that there is a relationship between the size of the brain and the size of the social group. When brain expansion came to a halt around 100,000 years ago, the size of a social group reached its maximum at 150 people. Dunbar thinks our biggest significant social groups still number about 150 people [28l], and that is too large to maintain social cohesion in the traditional way.

"One of the most important ways that primate allies show their affection to each other is by grooming. Grooming not only gets rid of lice and other skin parasites, but it also is soothing. Primates turn grooming into a social currency that they can use to buy the favour of other primates. But grooming takes a lot of time, and the larger the group size, the more time primates spend grooming one another. Gelada baboons, for example, live on the savannas of Ethiopia in groups that average 110, and they have to spend twenty percent of their day grooming one another."

Dunbar estimates that [28l]:

"If we had to bond our groups of 150 the way primates do, by grooming alone, we would have to spend about 40 or 45 percent of our total daytime in grooming."

Such a large amount of time cannot be combined with finding food on the savannas; it would, expressed in today's language, cause a non-sustainable work-life balance. Hominids therefore needed a better way to bond, and Dunbar claims that better way was language.

Deacon argues that language evolved from simple to complex variants under social pressure. Martin Nowak and his colleagues come to the same conclusion, but through the mathematical modelling of language evolution [28p]. They began with 'sound languages', like apes speak, and gradually allowed the sound vocabulary to extend. But because the 'sound space' is limited, at a certain moment communication turned into confusion, because new calls began to 'sound like' existing ones. The solution to this trap was to string sounds together, into sequences or words. Thus when life became so complicated that sound-speaking led to a tower of Babel effect, word-speaking species became more fit for survival [28p]. Then history repeated itself. Words had to be stored in brains. But the storage capacity of brain memory is limited. There was a moment that 'syntaxing' words into sentences became more effective: syntax enabled the turnaround of a few hundred words into millions of expressions. Carl Zimmer summarises what the Evolution Project revealed [28p]:

> *"A syntax-free language beats out syntax when there are only a few events that have to be described. But above a certain threshold of complexity, syntax became more successful. When a lot of things are happening, and a lot of people or animals are involved, speaking in sentences wins ... Something about the life of our ancestors became complex and created a demand for a complex way in which they could express themselves.*
>
> *A strong candidate for that complexity, as Dunbar and others have shown, was the evolving social life of hominids."*

Let us go back now to the pressing question posed at the beginning of this section: what on earth is the advantage that humans have over animals? The preceding interpretation of scientific thoughts represents the 'mainstream answer': the human advantage is the result of an extraordinary social evolution, or the capacity to introduce and grow complexity in social life. This, in turn, induced the co-evolution of brainpower and technique; with 'technique' meaning *"a method of accomplishing a desired aim"*.[*] Language, so considered, is a technique to communicate; early-stone-tool-use is a technique to prepare food, while painting is a technique to make visual art. It seems to be agreed that 'introducing complexity into social life' is the driver of human evolution, a self-propelling driver.[†]

But nothing is self-propelling. Modern energy theory says that complexity never ever drives anything. Complexity emerges and the emergence of complexity is driven by a flow of energy, without exception. In an evolutionary energetics approach, phenomena such as anatomy, stone technology, language and social structure are *emergent properties from energy systems, or energy economies*. Thus inspired by energetics, I read the history of life with other eyes or rather with another mindset.

---

[*] From Merriam Webster's Collegiate Dictionary, 10th Edition, 2002.

[†] Deacon believes symbol use itself must have been the prime mover for hominid brain evolution. Language has given rise to a brain, and not *vice versa* [56l]. At the same time he infers that symbolic referencing is intrinsically social [56l].

MAN THE FIRE MASTER

The human advantage

To find an evolutionary energeticist's answer to the quest for the human advantage, we will look at the same observations and discoveries as presented in the previous section but add the role of energy to the discussion:

- About 500,000 years ago human evolution gained momentum – *"the glacial monotomy of life began to break up"*;
- Somewhere between 600,000 and 500,000 years ago *Homo ergaster*, equipped with a medium-sized brain, left the field and *Homo heidelbergensis* entered the arena with a near-modern-sized brain;
- Perhaps somewhere between 500,000 and 150,000 years ago human language reached its modern form; building in the capacity to master language required a radical re-engineering of the brain;
- Unlike any other animal, humans developed the capacity to symbolise; speaking and thinking require symbolic referencing;
- Humans created advantage over other animals through extraordinary social evolution, or the capacity to complexify social life, which in turn induced the co-evolution of brainpower and technique;
- Starting 500,000 years ago humans learned how to keep controlled fires [28l,54n].

The last observation about fire control is often mentioned in chronicles about the history of human civilisation, but usually only as a casual remark. Hardly anybody *does* anything with this observation, the focus is on stone tools, body structure, language, hunting, and gathering habits.

One of the exceptions is the archaeologist Avraham Ronen who in 1998 issued the paper entitled 'Domestic fire as evidence for language'. In comparing communication between apes and humans, more specifically fire-controlling humans, Ronen concludes [57]:

> *"Thus ape cooperation forms most frequently for immediate purposes, involves only a few individuals, and is based on the evaluation of given and received behaviour like grooming, sex, and food sharing. On the other hand, the human coalition involved in keeping fire must have included the whole group and lasted for a long time, while the burden of supplying fuel brought only delayed benefits. This type of group activity appears to be peculiar to humans alone, and seems to indicate a type of social organization unseen among non-humans."*

The Pyro-Energy Revolution

Fire control demanded advanced communication. It created the need for communicable references to the environment, and therefore people invented names and abstractions, or symbols.

> *"Keeping and feeding fire must have involved the notion of object permanence and the notion of disappearance, as well as that of success and failure. Successfully keeping fire required testing cause and effect relationships, learning from past experience and planning, instructing prohibition, especially to children, and systematically using units of measurement (fuel quantity), hence, mathematical abstraction. Providing for the fire requires yet another feature which is absent in all communication systems but the human: displacement. Displacement means a message with no immediate sensory contact with the event to which the message refers, as is the case of fuel which is only indirectly related to the fire."*

Ronen connects fire mastery with the need for human communication based on symbolic referencing. Moreover, beyond being a tool, Ronen argues, fire became a symbol itself.

> *"Fire must have come to symbolize its most obvious representation: human singularity. Humans stand alone among the myriad of creatures around in possessing fire, a fact which could not have passed unnoticed. It is the ultimate weapon which no other can use or defend itself against. If there had been a trigger to arouse self-consciousness and the ultimate sense of "otherness", it was fire. Given the high cognitive capacity of Acheulean humans, it is unlikely that these humans were unaware of their difference from other animals and their unique position in the world. Symbolic processes express meaning based upon cognitive processes, and it is suggested here that in the course of the Lower Paleolithic, domestic fire could have come to symbolize the unbridgeable gap separating humans from animals, and the human condition as a whole. Through language, then, fire "became paradigm for all of humanity's interaction with nature" [from Stephen Pyne]."**

Another eminent scholar swimming against the mainstream is the sociologist Goudsblom, a rather special figure in the human history debate. In 1992 he published 'Fire and Civilization', a profound sociological study into the origins of human dominance. The development of tools and language dominate the human-animal dispute, but animals use tools and animals communicate through sounds, although much less

---

*Ronen states that the precise time of the domestication of fire is still debated, but there is good evidence that it was during the Lower Paleolithic [57].

advanced than we do. Is there something then that we do and they do not? Yes there is, contends Goudsblom [15a]:

> *"The ability to handle fire is a universal human attainment, found in every known society. It is also, to an even greater extent than either language or the use of tools, exclusively human. Rudimentary forms of language and tool use are also found among non-human primates and other animals; but only humans have learned, as part of their culture, to control fire."*

The third 'dissident' scholar I would like to introduce is the environmental historian Stephen Pyne. Pyne dedicated part of his life to the 'Cycle of Fire', a suite of six books that collectively narrate the story of how fire and humanity interacted to shape the Earth [49]. The foreword to 'World Fire: The Culture of Fire on Earth' reads [58a]:

> *"The story of fire goes so far back in the human past that one can plausibly view it – Prometheus-like – as the hearth from which all culture and civilization sprang."*

When humans appeared they experienced fire in the same way as rain and snow, or heat and cold, as an event that just came about [15b]. They possibly basked in the glow of smouldering embers and picked up partly charred animals and fruit from the ashes – as living animals still do – to enhance foraging; some uneatable fruits became edible, and storage life of meat improved. Such activities probably define the first and most long-lasting period of human fire use, which was opportunistic and predominantly passive – passive because humans did not actively collect and preserve fire.

For hundreds of millions of years natural circumstances caused fires to start and to stop, until hominids began to interfere: at some time in the past they began to stop the 'stopping' of wild fire. After following fires wherever they occurred, people learned how to keep a fire going at its original site, and eventually to transport it to safe and sheltered dwelling places [15b]. The fossil record of human artefacts does not include many 'fire artefacts'; they disappear 'by design'. It is very difficult, therefore, to mark the transition from passive to active use of fire. According to recent insights, humans learned how to make reliable fires less than 500,000 years ago [281,54n]. In those days, *Homo heidelbergensis* colonised Africa and later Europe, while *Homo erectus* inhabited China and South-East Asia; both are probably descendants from *Homo ergaster* [54g]. The logical argument for dating active fire use between 500,000 and 300,000 years ago seems particularly strong for the north Chinese *Homo erectus* and the European *Homo heidelbergensis*, *"both of whom occupied environments where fire would have been far more than a luxury"*, argues Richard Klein [54n]; but:

The Pyro-Energy Revolution

"If in fact, we insist on well-defined fossil hearths, the oldest firm evidence for human mastery of fire comes only from African and Eurasian cave sites that are younger than 250,000 years."

The origin of fire

In the beginning there was no fire, as some essential ingredients were lacking.

Approximately two billion years ago, the concentration of atmospheric oxygen reached one percent of its present value [45a]. That seems low, but oxygen shocked the Earth with its reactivity as we have seen in the preceding chapter [59a]. One of the marvellous triumphs of the following Aerobic Regime was the advent of land plants about 500 million years before the present day [28e]. Huge structures evolved through innovations in the plant's cell walls. Secondary walls, consisting of polymeric carbohydrates, such as cellulose and lignin, provided plant stability [34f]. Watch the double role of carbohydrates: solar energy storage and bio-construction.

Once ignited, with sufficient oxygen available, a plant's stored solar energy is released as heat and light through rapid and complete combustion. Water and carbon dioxide, once the building blocks in carbohydrate photosynthesis, now become degradation products. If the heat of combustion maintains the ignition temperature, combustion sustains itself and there is fire!

When the green plants arrived on land, the concentration of atmospheric oxygen was about 15 percent [45a]. Laboratory experiments suggest that above 12 percent a wild fire can start [59a]. Wild fire, however, does not start spontaneously. It needs to be ignited. An external factor must heat at least one spot on a plant up to its ignition temperature. A flash of lightning, for instance, or a volcanic eruption can do the job.

Approximately 370 million years ago the first fossilised charcoal appeared *"suggesting that oxygen had risen sufficiently to sustain fires"* [37f]. All the necessary ingredients were there: combustibles, in the form of plant matter; air, with atmospheric oxygen above 12 percent; and clouds, to enable ignition through lightning. Life had prepared the Earth for wild fire.

Active use of fire

The transition from passive to active use of fire is momentously important. It marks the first great act of human interference with natural processes, or the first hesitant beginnings of a fuel-driven economy [60a]. Through the domestication of fire, humans changed from an 'ecologically secondary' to an 'ecologically dominant' species [15b]. Stephen Pyne expresses the historic change as follows [49b]:

"In Swartkrans, a South African cave, the oldest deposits hold caches of bones, the prey of local carnivores. Those gnawed bones contain the abundant remains of ancient hominids. Above that record rests, like a crack of doom, a stratum of charcoal; and atop that burned break, the proportion of bones abruptly reverses. Above the charcoal, the prey have become predators. Hominids have claimed the cave, remade it with fire, and now rule. That, in a nutshell, is what has occurred throughout the Earth. What has happened with early prey relationships happened also with fire. As humans successfully challenged lightning for control over ignition, the whole world has become a hominid cave, illuminated, protected, nurtured, warmed, and controlled by the flame over which humanity exercises its unique power and through which it has sought an ethic to reconcile that power with responsibility. Ours became the dominant fire regimes of the planet."

Returning again to the fundamental question propounded at the beginning of this section: what on Earth is the advantage that humans have over animals? We have seen that the mainstream reply is: extraordinary social evolution. Let me now develop the perspective from an evolutionary energetics angle.

About 500,000 years ago human evolution gained momentum – *"the glacial monotomy of life began to break up"*. A revolutionary new energy economy, the fire economy, induced sweeping acceleration of social evolution under its originators, the fire masters. This, in turn, drove the co-evolution of the masters' brainpower and technical skills to continuously higher levels of complexity. The brain size of the early fire master, *Homo sapiens'* predecessor *Homo heidelbergensis* was nearly ours already but evolving fire control induced a radical re-engineering of his whole brain speeding up the evolution of his capacity to learn symbolically, speak, and think. Phenomena such as language, consciousness, social coordination, cultural transfer and imaginative power arose from the new energy economy. Maybe the most radical 'emergent property' was the fire masters' rising ecological dominance. The fire master was a fairly good toolmaker and he became a very good talker, but the true human advantage was fire mastery. 'Man the Fire Master' made the human difference. Stephen Pyne [49c]:

"According to many myths, we became truly human only when we acquired fire."

The Pyro-Energy Revolution

The Pyrocultural Regime

Fire mastery is a rudimentary form of energy mastery and, in the light of the history of life, a truly out of the ordinary manifestation of it. The preceding chapters discussed how more than 3.5 billion years ago innovative microbes developed photosynthesis: they harvested sunlight and stored it chemically in carbohydrates. Then, around two billion years ago, a successful microsymbiotic alliance integrated photosynthesis with aerobic respiration. During the Photoic Era living organisms gained energy through in-cell, or *in vivo*, combustion of crude sugars; first, anaerobically in the Photian Period, and then, later, aerobically in the Oxian Period. But about half a million years ago the energy economy changed fundamentally with the introduction of pyrotechnological innovations. On the time scale of the history of life, half a million years in 3.5 billion years is akin to half a second in an hour.

In vivo combustion of natural carbohydrates is only applicable to digestible fuel, which we call food. Yet half a million years ago, just like today, by far the largest resource base of photosynthetic carbohydrates was indigestible; humans cannot digest lignin, cellulose, or hemicellulose, the major parts of wood [33d]. But inedible does not mean incombustible; inedible only means incombustible through the use of cellular 'internal combustion technology'. We all know that carbohydrates packed together in wood burn fiercely. And when they burn, the stored solar energy is released rapidly in the form of light and heat. Humans invented ways to utilise the solar energy stored in non-digestibles for improving the quality of their life. Fire technology made a virtually endless energy resource base instantly available to its masters.

Perhaps the most profound implication of fire mastery was that for the first time – after maybe 3.9 billion years of evolution – the genes lost their monopoly in 'energy mastery'. For aeons they had controlled the flow of energy through every living creature. Look at us. We eat, but after swallowing a tasty morsel we leave the further processing to the countless mitochondria in our body. These micro-plants are placed under gene government, so the digestion of the food we consume is a completely *subconscious* affair. In contrast, the 'digestion' of wood through fire control is a *conscious* and *collective* process. The emergence of fire mastery was not only brain-controlled – a revolution itself in the eyes of the genes – but it was a group of brains that took control. Fire mastery is an outstanding example of an advanced social activity. And because this social activity endowed its practitioners with an extraordinary competitive advantage, social evolution accelerated. In terms of energetics, the advent of fire mastery was so radically different from the biological modes of energy control in the Photoic that it defines a new energy era: the Pyroic. The Pyro-Energy Revolution that brought fire mastery was a true solar energy revolution, the third one in succession, because the fire masters began to exploit a new source of solar energy for the development of their communities. In the form of wood, additional amounts – by orders of magnitude – of solar energy were made available to the fire masters. They established a new energy regime: the Pyrocultural Regime. Johan Goudsblom introduced the term Fire Regime to indicate a corresponding *"ecological regime"* [15c]. Through fire mastery the fire masters quickly gained ecological dominance.

First life tamed sunlight, then oxygen, and after that fire. But fire mastery was, and still is a monopoly held by only one species: genus *Homo*. Says Pyne [58b]:

"A fire creature came to dominate a fire planet."

Genus *Homo* was, and still is the only mammal on Earth defying the fear of fire. He transferred the power of fire from the cloud to the hand [59b]; with the torch he captured fire, and transported it; with the torch he stopped the 'stopping' of fire, and domesticated it. The torch became a universal tool, an enabling device for endless biotic technologies. It was about 9000 years ago before Neolithic man acquired reliable fire-making techniques. And even then it was more convenient to keep a fire alive permanently than to reignite it [49d]. What mattered was preservation, not ignition.

Keeping a fire going required foresight and care [15d]. Wood-fuel needed to be gathered, and perhaps even stored during wet periods. The fire had to be fed continuously, and to be fanned and bounded simultaneously. People brought fire under their roof, with children playing around. Such activities, which demanded attention and consultation, were not done instinctively, but required self-control instead. Fire mastery was not genetically determined, but had to be learned. Goudsblom regards the ability to know about fire and the readiness to care for it as a mental or psychological quality [15d]. These attributes complemented physical characteristics such as upright walking, flexible hands, and a large, differentiated brain. However, the physical and mental conditions of fire control could only develop in community. Only by passing on the acquired knowledge to the young, could people obtain permanent power over fire. Says Goudsblom:

"Both thought and cooperation were stimulated by the very efforts which the control of fire, as a technical problem, demanded. The technical problem was, at the same time, an intellectual and emotional problem, and a problem of social coordination."

The Pyro-Energy Revolution

The upkeep of technical and mental capacities necessitated a sociocultural framework [15d]: through *social coordination* one ensured, for example, that there was always someone to look after the fire; through *cultural transfer* one guaranteed the survival of the skills, as well as the sense of responsibility and duty with respect to fire use. Both social coordination and cultural transfer were necessary preconditions for the domestication of fire. But, in addition, these social functions were reinforced by the very domestication. A true co-evolutionary process began: people adapted fire to their needs, but had to adapt their habits to fire. In this sense, Johan Goudsblom infers, *"the domestication of fire also involved 'self-domestication' or 'civilization'."* Pyne agrees [58c]:

> *"But if fire granted early humans new power, it also conveyed responsibilities. It was vital that the flame neither fail nor run wild. Domestication thus began with the domestication of fire, and this in turn demanded the domestication of humans. Fire could not thrive unless it was tended, sheltered, fed, nurtured. The most basic social unit consisted of those people who shared a fireside. The hearth was the home."*

Societal metabolism

Fire mastery introduced a novel, *'ex vivo'* metabolic site, the fireside. The cellular mitochondria continued to take care of the fire masters' *internal* combustion of food, but the fireside added an extra energy supply through the *external* combustion of wood. Hence fire control created a whole new fuel economy. The new economy not only expanded energy utilisation radically, but also led to innovative energy chains: whereas cell metabolism delivers *bio*-chemical and *bio*-mechanical work, 'fire metabolism' yields heat, light, and *'anthropo*-chemical' work (such as roasting, and later cooking). Just like cell metabolism, fire metabolism processes green plant matter to utilise the stored solar energy. But fire metabolism marked the origin of a fundamentally different kind of metabolism, *societal metabolism*.*

Cell metabolism is the capability of a living cell to capture and transform matter and energy to fulfil its needs for survival, growth, and reproduction [39a]. Similarly, *societal metabolism* is the capability of a human society to

---

*Fischer-Kowalski and Hüttler describe societal metabolism as *"the whole of the materials and energy flows going through the industrial (and subsistence socio-economic) system(s)"* [61]. I generalise the concept of societal metabolism to the whole of the energy and materials flows going through human communities, or more precisely, the ability of a human community or society to capture and transform energy and matter from its environment in order to supply its needs for living, social coordination, and cultural transfer.

capture and transform matter and energy to fulfil its needs for survival, economic growth and cultural transfer. Cell metabolism is biogenic; societal metabolism anthropogenic. The 'execution code' for cell metabolism is programmed in the DNA, while the rules of societal metabolism are encoded in human brains and artefacts. As cell metabolism powers living organisms and thus biological evolution, societal metabolism drives human communities and thus cultural evolution. And just as cell metabolism creates biological diversity, societal metabolism generates cultural diversity.* A living organism needs a continuous flow of energy to stay alive, and a society in which people *live together* needs a continuous flow of energy to thrive; imagine that the energy flow through a living organism or a human society comes to a halt ... then the organism dies or the society collapses.

Both cell and societal metabolism provide sustenance thanks to two fuel economies, 'food-fuel' and 'wood-fuel', respectively. These fuel economies became linked in a co-evolutionary process. According to the French archaeologist Catherine Perlès, the impact of 'the culinary act' was immense [15e]. Cooking induced physiological, psychological and social change: physiological as a result of extending the diet with substances that would otherwise not have been readily digestible; and social and psychological because people started to cultivate eating habits of their own, by which they could distinguish themselves from other animals as well as from each other. In terms of fire metabolism, cooking is using heat from fire to perform chemical work in raw food with the aim of changing the molecular composition; the heat, as we have seen, is released by combustion of the carbohydrates in the wood-fuel, a storage reservoir for solar energy.

The Symbolisational Signal

Paleoanthropologists refer to the human species, as it evolved in Africa between 200,000 and 50,000 years ago, with the qualification 'near-human' [28q,54o]. Then modern humans, or *Homo sapiens* emerged. Again, as had happened with *Homo ergaster* and *Homo heidelbergensis*, the arrival of a new human species coincided with a pronounced discontinuity in the development of stone tools: the transition to the Later Stone Age; blades narrowed and styles changed rapidly [55e]. However, focussing on stone

---

*Cultural diversity is social plus mental diversity, with mental diversity understood as the whole of consciousness, conscience and intelligence.

The Pyro-Energy Revolution

tools alone may give a wrong impression of what actually occurred. *Homo sapiens* appeared to be cosmopolitan [28r]:

> "Only around 50,000 years ago did they sweep out of Africa, and in a matter of a few thousand years they replaced all other species of humans across the Old World. These new Africans did not just look like us; now they acted like us. They invented tools far more sophisticated than those of their ancestors – hafted spears and spear-throwers, needles for making clothes, awls and nets – which they made from new materials like ivory, shells, and bones. They built themselves houses and adorned themselves with jewellery and carved sculptures and painted caves and cliff walls."

It sounds spectacular that the European modernisation was simply an outgrowth of behavioural change that occurred in Africa perhaps just 5000 years earlier [54p]. And it happened only yesterday [28s]:

> "In an evolutionary flash, every major continent except for Antarctica was home to *Homo sapiens*. What had once been a minor subspecies of chimp, an exile from the forests, had taken over the world."

But apart from more 'useful' activities such as effective hunting, building solid houses, making tailored clothes and improving fireplaces [54o], modern humans began to create 'non-useful' artefacts in the form of art, jewellery, ornaments and musical instruments; the oldest clear musical instruments are 30,000 years old bird bone flutes [54q]. Also, 50,000 years ago, elaborate graves with unequivocal ideological or religious implications show up in the archaeological record [54r]. All these 'paleo-signs' indicate the origin of ethnographic 'cultures' or identity-conscious ethnic groups [54s]. Interestingly the rise of rites is related to the domestication of fire [15f]:

> "Since no hominid or human group could afford to start from scratch and invent for each new generation new solutions to the problems posed by keeping a fire, every group had to rely largely upon socially standardized procedures, or rites. Rites, as learned alternatives to "instincts", have continued to be attached to the use of fire in every known society to the present day. As it is most likely that this has been the case ever since hominids first learned to keep a fire going, it stands to reason that the control of fire has also contributed to the development of the general human capacity to engage in ritual."

Maybe language came to its fullest flower during the cultural climax [28t]. The making of symbolic representations – such as a painting, or an ornament – required an advanced capacity to symbolise, or well-developed powers of imagination. It is conceivable that artificial symbols could only emerge in communities of people with shared values and beliefs [55e]. The archaeological annals expose a sudden growth of symbolic activity with the arrival of *Homo sapiens* 50,000 to 40,000 years ago.

Several scholars emphasise that modernity came as a real bombshell, and speak of the *"dawn of human culture"*, a *"human revolution"*, a *"creative explosion"*, a *"great leap forward"* or a *"socio-cultural big bang"* [54t].* I see this step-change in human awareness as a Symbolisational Signal indicating that human beings adopted a new perception of reality. A new way of thinking allowed people to think abstractly about nature and themselves [28t]:

> *"The artefacts that humans left behind speak to a profound shift in the way humans saw themselves and the world. And that shift may have given them a competitive edge."*

The uniquely human capacity to symbolise evolved in a social group of fire masters. The Symbolisational Signal, just like the corresponding new perception of reality, is an emergent property of the fire economy, the energy regime based on fire mastery. Fire mastery brought genus *Homo* the almost infinite ability to innovate. It markedly accelerated the evolution of social learning and thus the co-evolution of brainpower and technique. Brainpower and technological innovation have been key differentiators of human culture since its very origin around 2.5 million years ago. And through fire culture† genus *Homo* established ecological dominance.

AND THE FACE OF THE EARTH CHANGED

A changing lifestyle

Fire mastery equipped people with astonishing power. Listen to Pyne [59b]:

> *"Fire shielded hominids from the cold, it pushed back the night, it warned off predators, it extended the hunter's range and the forager's*

---

*Not all archaeologists agree. Some find that the behavioural differences between the Middle and the Later Stone Age are exaggerated, arguing that the real advance to modern behaviour occurred with the appearance of the Middle Stone Age, about 250,000 years ago. See Richard Klein [54u].

†The term 'culture', and thus 'human culture', has many definitions. Here I follow Richard Klein who states that *"The first* [trigger] *event* [in human evolution] *occurred around 2.5 million years ago, when flaked stone tools made their initial appearance. These comprise the earliest enduring evidence for human culture."* [54d]

grasp – driving and attracting grazers, promoting and extinguishing plants. It assisted in tool-making by hardening wooden spears, quarrying stone, glazing flint; the techniques of drilling, scraping, and striking are the same for making fire as for shaping tools from bone, wood, and stone; tools and torch reacted synergistically, each amplifying the power of the other. Hunting, felling, and gathering altered the array of fuels, making fire more or less potent, while controlled burning greatly magnified the ability to fell trees, clear shrubs, stimulate grasses, drive or draw fauna, and select for or against resident flora. For nearly all environments, fire provided a means of access, and for some, a medium of rapid domination."

Simple foragers evolved into sophisticated hunter-gatherers and incipient plant cultivators [33e]. In using fire for hunting, they changed the land they inhabited – at first perhaps inadvertently, and later deliberately [15g]. Probably fires were lit intentionally to drive game animals out of their lairs in the bush. But after a 'hunting fire' new vegetation, such as grasses and legumes, arose spontaneously, and these in turn attracted grazers. Humans observed well, and began to create ecological circumstances beneficial for themselves. In fact they invented fire technologies to cultivate land. Moreover, land cultivation encouraged people to make it their own and establish 'dominion' over it.

Through adding the force of fire to their own, fire masters drastically increased their productivity. Fire technologies facilitated the impressive territorial expansion of *Homo sapiens* over the globe [15e]. Dark evenings could be filled with labour, play and ritual. Fire also provided people with security, as it kept predators and other animals at bay:

"Fire acted as a buffer against the extremes of cold and darkness, creating small enclaves of heat and light."

More effective hunting methods, cooking, warmth and light facilitated both 'intensive growth', that is lifestyle innovations, and 'extensive growth', which means population increase. Intensive growth is harder to define and measure than extensive growth; moreover, the two phenomena can support and reinforce, or obstruct each other. Intensive growth usually leads to a shift in existing balances of power as well as to changes in mentality or habits [15g]. Fire technology gave humans a competitive advantage over both hominid and non-hominid competitors. In doing so, humans increased the carrying capacity of the Earth for their own population. But also, fire mastery was so successful that, as an unplanned side effect, the masters became increasingly vulnerable to losing it [15f].

And life changed the land

Many of the ecological and cultural effects due to fire mastery were wiped out by the drastic climate changes during the late Pleistocene, or the Great Ice Age, which reached its peak between 22,000 and 16,000 years ago. But soon after the ice caps had disappeared from what became temperate zones where trees could grow again, humans began applying their ancient burning practices to the new virgin woods. Archaeologists have found evidence of incipient deforestation caused by anthropogenic fires in many places [15g]. Says Stephen Pyne [58d]:

> *"Anthropogenic fire reshaped the structure and composition of landscapes, recalibrated their dynamics, reset their timings of growth and decay. Human's ability to manipulate fuels redesigned the environment within which fire – either theirs or nature's – had to operate."*

With an increasing population the fire economy grew [15g]. Gradually the relationship between the fire master and his environment intensified, becoming more intricate. Then, most probably triggered by a series of concurrent factors, both from natural and cultural origin, a next energy dominancy emerged from the Pyrocultural Regime. The Pyrian period was the first period of the Pyroic Era, indeed, but not the last.

– 4 –
The Agro-Energy Revolution

MAN THE SOLAR FARMER

The Agricultural Revolution

In terms of evolutionary energetics, *Homo sapiens* appeared on the Earth's landscape in the Pyrian, the first period of the Pyroic Era, when pyroculture emerged and flourished. In geological terms, our forefathers were born in the final stage of the Pleistocene, or the Great Ice Age, which lasted until 10,000 years ago [62a]. Then the Holocene began, the epoch that we live in today. Great environmental transitions marked the changeover from Pleistocene to Holocene [63a]:

> *"As the ice melted and the glaciers receded, the sea level rose world-wide by at least 100 metres, terminating the land bridge between Siberia and Alaska and turning large sections of the Eurasian continent into islands, including the British Isles and Indonesia. As the temperature rose, the tree line shifted away from the equator, turning tundra and savannah into woodland and forest."*

During the Pyrian period most hunter-gatherers spent more than nine-tenths of their time on hunting and gathering [33e]. Although the roaming foragers increased the size and range of their habitat tremendously, population densities were hardly higher than those of their primate predecessors. The least hospitable environments, such as tundras and boreal forests, supported densities of just one person per 100 squared kilometres while the most suitable habitats, such as the tropical and temperate grasslands, could have a carrying capacity of one person per squared kilometre [33f].

Period	Age
Agrian	About 12,000 years ago
Pyrian	About 0.5 million years ago
Oxian	About 2.1 billion years ago
Photian	About 3.8 billion years ago
Thermian	About 4.2 billion years ago

These environments offered better biomass supply, a higher accessibility and edibility of plants, together with more and bigger animals for hunting.

Hunter-gatherers gradually familiarised themselves with the behaviour of grazing animals and edible crops. This experience stimulated two fairly different ways of living: nomadic pastoralism, or livestock raising, and shifting cultivation, or crop raising. Nomadic pastoralism did not lead to significant population growth – farm animals are relatively inefficient converters of plants to flesh – but shifting cultivation certainly did [33f]. Early agriculture involved alternation between short periods of cropping – commonly just one season, rarely more than three years – and long spans of fallow, lasting at least a decade. The cropping cycle often began with partial removal of natural overgrowth, accomplished in forests by 'slash and burn' and on grasslands simply by setting fires. The net energy returns of shifting cultivation were impressive: growing grains yielded a factor of 10 to 15, which means that the energy output was 10 to 15 times higher than the energy input, mere manual labour in those days. Even less physical effort was needed for growing corn; energy returns were commonly more than 20-fold. And for tropical tubers, legumes and bananas, the best energy profits even ranged between factors of 40 and 70 [33f].

The Fire Regime brought forth the first great transition in human civilisation: the Agricultural Revolution. Where fire had firstly *"opened up the land for hunting"*, it now prepared the land for cultivation [62b].

The forces that shaped the revolution

Unravelling the shaping forces in the gradual transformation from pyro- to agriculture – or 'agrarianisation' (like industrialisation) as it is termed – is not easy [62a]. No single vector, whether climatic, biogeographical or sociopolitical by nature, displays a preponderant correlation with the rise of agriculture [62c]. It seems to have been a real co-evolutionary process between the developments of the earthly environment and human culture.

Agriculture came to prominence in the transition period between the last glacial period, the Pleistocene, and the current interglacial, or Holocene age – so it arose in a time of serious climate change.

> *"The last glacial period, from 22,000 to 13,000 [years ago], was very cold and dry throughout Europe ... Forest and woodland would have been almost non-existent ... The majority of the area was characterized by a sparse grassland or semi-desert coverage. Following initial warming, as the ice mass started to melt, some open woodland cover appeared quite rapidly."* [64]

The Agro-Energy Revolution

Nowadays the prevailing view is that agriculture originated some 12,000 years ago in the Levantine Corridor [62a], *"the narrow band of habitable land in Israel and Jordan that offered the most obvious way for the ancestors of modern humans to disperse out of Africa"* [65]. In the beginning people domesticated cereals and pulses, and within a millenium animal domestication took place: first dogs, then sheep and goats [62a]. Presumably by this time the steppe vegetation had started to become richer across the whole Fertile Crescent with its alluvial plains near three 'Great Rivers' – the Euphrates and Tigris in Mesopotamia and the Nile in Egypt [62d].

Food production became the core of the human/environment interaction. But the human/environment system showed positive and negative feedback loops: climate change caused alterations in vegetation; these then improved natural food supply and thus increased the carrying capacity of the land; this led to sedentism and higher population densities, which in turn boosted the food-quest and forced people to begin crop cultivation – agriculture, it is thought, was *"'progress' born out of necessity"*, enabled by both technological innovations in the form of more advanced tools and practices, and 'societal innovations' such as better organisational structures and trade [62a,e]. But also, climate change caused micropredators to spread; and because higher population densities increased people's vulnerability to germs, higher mortalities could occur [62f]. Note that (with just a few exceptions) high-density populations did not develop in tropical river basins with their usually much larger water streams, probably because of *"disease-causing micropredators and poor soils"* [62d].

Vegetation change induced by climate change kept influencing the habitability of land – *"the melting of large ice sheets in Europe made the large Eurasian steppes and the northern lands of Europe and Canada available for habitation"* [64]. A great variety of the so-called traditional agricultures emerged [33f], leading to bigger harvests either by the extension of arable land or by the gradual intensification of cropping, the latter resting essentially on three advances:

- Greater use of draught animals, which eliminated much of the heaviest human labour;
- Irrigation and fertilisation, which improved the supply of the two inputs – water and minerals – most often limiting crop productivity;
- Cultivation of a greater variety of plants, which made cropping not only more productive but also more resilient.

Through innovating the food economy traditional farming succeeded in supporting population densities an order of magnitude higher than its agricultural antecedent, shifting cultivation. Remarkably, despite the use of animals, human physical exertion remained the dominant force, also for the sophisticated technologies such as sowing, hoeing, weeding, and harvesting, until as late as the 19th century [33f]. Whereas domesticated cattle and horses helped with threshing and lifting water for irrigation, deep ploughing in heavy soils was their most indispensable service [66a].

Agriculture emerged in a polygon of shaping forces. In 'Mappae Mundi: Humans and their Habitats in a Long-Term Socio-Ecological Perspective', Bert de Vries and Robert Marchant maintain that agrarianisation *"was more of a gradual intensification of the relationship between groups of humans, their environment and each other"* [62g]. And Johan Goudsblom, in painting a sociological perspective, keeps us from committing environmental determinism [60b]:

> *"Religious-agrarian regimes have played an important part in shaping the relations between humans and the biosphere in agrarian societies. Humans are not equipped by birth with a natural aptitude for agrarian life. They have no innate calendar telling them when the time has come for preparing the soil, for planting the seeds, for removing weeds, for harvesting. The only calendar available to them is a socio-cultural one, roughly geared to the alteration of seasons but regulated with greater precision by human convention, as a part of an agrarian regime."*

The Agrocultural Regime

The Agrian Period experienced an enormous increase in the carrying capacity of *Homo sapiens'* habitat; from just 0.01 to 1 person per squared kilometre with foraging to 10 to 15 in the case of shifting cultivation, and up to about 100 to 800 people per squared kilometre when people started practising traditional farming [33f]. That represents an upsurge in food yield of 3 to 4 orders of magnitude. Still, the farmers did not tap a new source of energy, nor did they invent more efficient combustion technology. In essence, agricultural innovation involved the economically successful domestication of edible plants and grazing animals. It amounted to switching from *"wild plant-food procurement to crop production"* and from *"predation to taming and protective herding to livestock raising and pastoralism"* [62a].

When viewed from an evolutionary energetics angle, *food* production is equivalent to *fuel* production, because food fuelled the muscles, and muscles were the prime movers in the agricultural economy. Hence the Agricultural Revolution was a true energy revolution, the Agro-Energy Revolution. Because cultivating food involves harvesting sunlight, the

The Agro-Energy Revolution

Agro-Energy Revolution was a solar energy revolution, and the ruler of the new energy regime a 'solar farmer' [33e]. Solar farming enabled human beings to revolutionarily increase the utilisation of solar energy for the development of their societies, which led, in time, to the establishment of a new ecologically dominant societal metabolism, the Agrocultural Regime.*

The Agro-Energy Revolution is the fourth successive solar energy revolution in the history of life: the first was centred around the taming of sunlight, the second of oxygen, the third of wildfire, and the fourth of land. The solar farmer built on the successes of fire mastery, just as the aerobes exploited the photosynthesis of blue-greens – the energy regimes are evolutionarily connected. With the Agro-Energy Revolution a new energy period began, the Agrian. It is the second energy period in the Pyroic Era, because fire control remained a crucial foundation for land cultivation.

The emergence of competing energy chains

By using draught animals, the solar farmers augmented their societal metabolism with a new energy chain: they fed domesticated animals with self-produced food-fuel. The animals transformed the farmed solar energy into biomechanical work, which was subsequently converted into mechanical work through the coupling with a machine, such as a plough.

Through metallurgy, human beings expanded their societal metabolism yet again. The early metallurgy was very energy intensive; it consumed large amounts of wood, or charcoal made from wood. The high wood demand of medieval and early modern iron-smelting created many deforested landscapes in ore-rich regions [33g].

Its origin is not known precisely, but the history of waterwheels is generally assumed to have started in the first century BC, initially in Greece, then in Rome. This was an enormous event in the annals of solar energy harvesting as a waterwheel captures energy from a water stream without 'consuming' fuel. Because the sun sustains the climate, and the climate system maintains the world's water streams through evaporation and rain, a water stream is actually sun-driven. The waterwheel enabled humans to expand their societal metabolism with a renewable energy chain for the production of mechanical work without the involvement of photosynthesis or muscle power. In its relatively short history the

---

*I have chosen to use the term Agrocultural Regime instead of Agricultural Regime so as to stay consistent with Pyrocultural Regime. Obviously I will not flaunt conventions and keep talking about the Agricultural Revolution, which corresponds to the term Agro-Energy Revolution in the consistently developed nomenclature of the present work.

waterwheel innovation developed into the prime mover of the pre-industrial West [33h].

The windmill became common in the 12th century. By harvesting solar energy from the air [33h], the windmill was complementary to the water mill until the early 19th century. Another energy chain based on wind-harvesting – sailing – for centuries provided the only means of propulsion available for long-distance waterborne transportation [33i]. At their peak the sleek, fast sail ships developed into icons of growing international trade, and the heavily armed battleships became instruments of global empire building. After 1500 AD warships were equipped with guns representing yet another man-made energy chain augmenting *Homo sapiens'* societal metabolism.* As an end-use device, a gun converts rapidly released energy from ignited gunpowder into the fast movement of a bullet. Those who mastered this energy conversion best gained evolutionary advantage.

A NEW REALITY EMERGES

The Quantificational Signal

In his fascinating book 'The Measure of Reality: Quantification and Western Society, 1250 to 1600', the historian Alfred Crosby argues that near the end of the Agrocultural Regime, and the birth of Renaissance in Western Europe, during the Middle Ages, a truly new quantitative model of reality began to displace the ancient qualitative model. To grasp the depths of this *"perception revolution"* we need to understand how people thought before, what Crosby calls, 'The New Model' emerged. Therefore he goes back to the writings of Plato and Aristotle [16a]:

> *"These two men thought more highly of human reason than we do, but they did not believe our five senses capable of accurate measurement of nature."*

For example, Plato decided that the number of citizens in the ideal state was 5040. Not because, to hazard an explanation, a gathering of people of that size could hear one individual speaker without special amplification, but because 5040 is the product of $7 \times 6 \times 5 \times 4 \times 3 \times 2 \times 1$. According to Crosby, up to the early Middle Ages, Westerners paid little attention to the concept of 'measurable reality' (the master masons of the Gothic cathedrals

---

*Gunpowder had already existed for a few centuries. It originated in the 9th century in China and made its way west in the 13th century. The first explosive was black powder, a mixture of saltpetre (potassium nitrate), sulphur and charcoal. The recipe was further refined and finally fixed in the 14th century (from "gunpowder" Britannica Concise 2004).

were an exception). In the period between 1250 and 1350, however, a marked shift in people's 'quantification attitude' occurred. They began to think in quanta time, space and value. The mechanical clock, marine charts, perspective painting, double-entry bookkeeping, and note length in musical notation, came to light. Money as an abstract measurement of 'worth' appeared [16b]. Price began to quantify everything, including time, which was quantifiable in equal hours since the invention of the clock. *"In the dizzy vortex of a cash economy the West learned the habits of quantification"*, reasons Crosby. In effect, he adds, a swelling Quantificational Signal was about everywhere [16a]. And the New Model initialised revolutionary change, deeply touching human's perception of reality.

The Scientific Revolution

The shift in perception of time and space brought Nicole Oresme (circa 1325 to 1383) to the conclusion *"that reason did not provide the means to decide the heavens or the earth was turning"* [16c]. Copernicus (1473 to 1543) turned the universe of the ancient Greeks inside out, underpinning his extravagant conclusions with arguments similar to those of Oresme, but with added mathematical theory. It is common now to date the beginning of the so-called Copernican Revolution, or Scientific Revolution, to 1543 when two pivotal texts appeared. The first is Copernicus' 'De revolutionus orbium coelestium' (On the Revolutions of the Heavenly Spheres), and the second Andreas Vesalius' 'De humani corporis fabrica' (On the Fabric of the Human Body). The Scientific Revolution changed the world from *"the world of the more-or-less"* to *"the world of precision"* [67a].* With the invention of spectacles around 1300, sight became the first sense aided by technological devices [16d]. Then, three centuries after its origin, in 1608, the Dutch spectacle-maker Zacharias Jansen built the first operational microscope [68]. According to Hans-Joachim Schellnhuber, Jansen's creation was a turning point in scientific history: *"For the first time the human eye could transcend its natural limits and begin to explore the wonders of the microcosmos."* Only one year later, in 1609, Galileo Galilei constructed the first telescope, opening the macro-cosmos for detailed exploration. The microscope and the telescope unleashed a real revolution in observation.

---

*In 'Van Spierkracht tot Warmtedood: Een geschiedenis van de energie' (2002), M.J. Sparnaay refers to H.F. Cohen, who states in 'The Scientific Revolution' (1994) that Alexander Koryé, historian of science, in 1948 discriminated between the world before circa 1600 and the world after. The quotes *"the world of the more-or-less"* and *"the world of precision"* refer to Koryé [67a].

The co-evolution of 'new thinking' and 'new observing' gave wings to reductionism – *"the study of the world as an assemblage of physical parts that can be broken apart and analyzed separately"* – as it was introduced by René Descartes in works published over the years 1637 to 1649 [69a]. According to Edward O. Wilson, *"reductionism and analytic mathematical modeling were destined to become the most powerful intellectual instruments of modern science"*. Let me give one dramatic example of applied reductionism: the changing nature of fire. During the 17th century the so-called phlogiston theory had taken root, according to which theorem combustion of matter liberated the invisible material phlogiston [15h]. However, *quantitative* experiments revealed that combustion yielded higher weights, rather than lower ones. When Antoine Lavoisier discovered that during combustion matter combines with oxygen, the phlogiston doctrine died [59c]:

> *"Anthropogenic fire became a focus for understanding: it did for knowledge what it did for wildlands and dwellings. Fire was idea, symbol, subject, and tool. It could rework thought as it did metal or clay ... Within it gods were manifest, about it myths were told, by it philosophy was explored, and out of it a science evolved that would, in the end, destroy fire's magic, mystery, and metaphysic."*

For centuries scientists were driven by an inner urge to unravel the natural order. They had no practical applications in mind. In 1751, the famous 'Encyclopédie' of Diderot and d'Alembert, one of the principal works of the Age of Enlightenment, did not include the header 'technique', while the entry 'art' exposed a remarkable hint to scholars:*

> *"From way back one discriminates between the* Artes Liberales *and the* Artes Mechanicae. *The free arts are especially the work of the mind, while the others are especially handiwork. This distinction caused respectable people to be not interested in the practical application of experimental knowledge. They view attention for practice and the mechanical arts an activity the human mind unworthy."*

In roughly 12,000 years of agricultural history, brainpower and technique co-evolved rather significantly, though apparently not to the full of Diderot's and d'Alembert's imagination.

> *"Early in the 17th century, the natural philosopher Francis Bacon ... had advocated experimental science as a means of enlarging man's dominion over*

---

*Translated from Dutch; source *Techniek als cultuurverschijnsel*, Dutch Open University, 1996.

> *nature. By emphasizing a practical role for science in this way, Bacon implied a harmonization of science and technology, and he made his intention explicit by urging scientists to study the methods of craftsmen and craftsmen to learn more science. Bacon, with Descartes and other contemporaries, for the first time saw man becoming the master of nature, and a convergence between the traditional pursuits of science and technology was to be the way by which such mastery could be achieved.*
>
> *Yet the wedding of science and technology proposed by Bacon was not soon consummated. Over the next 200 years, carpenters and mechanics – practical men of long standing – built iron bridges, steam engines, and textile machinery without much reference to scientific principles, while scientists – still amateurs – pursued their investigations in a haphazard manner."* [70]

It took time, but in the end scientists and craftsmen forged a symbiotic partnership. The new alliance, termed 'modern technology', cultivated reductionism and physical law. It happened to unleash unequalled innovative power in the centuries to come, but – there is a drawback to everything! – it also induced the emergence of a science schism between the natural sciences and the cultural branches of *arts* and *humanoria*, as mentioned already in the introductory chapter.

AND THE FACE OF THE EARTH CHANGED

From farms to nation states

During the Agrocultural Regime *Homo sapiens* changed the face of the Earth fairly radically. The human population expanded exponentially from about 4 million to roughly 900 million people around 1750 [66b]. Solar energy handling in the form of food production was the driving force. Vaclav Smil says [53b]:

> *"Plowing opened the soil for planting of small cereal seeds on scales vastly surpassing those of hoe-dependent farming. All of the Old World's high cultures were creations of grain surplus, and regular plowing was their energetic hallmark."*

The domestication of crops, and also of animals, allowed nomadic people to establish farms on their land. *Homo sapiens* settled down, an extraordinary event in human history. Communities grew steadily; settlements developed from farms into groups of farms and subsequently into villages. As long as communities stayed small, people needed hardly any government. The division between ruler and ruled existed, if at all, only within the family. And between families within the village, the status of every elder or family head was equal. But the rise of agroculture began to change this state of affairs, as the Encyclopædia Britannica, 2003, reads under the lemma 'government':

> *"In the land of Sumer (modern Iraq) the invention of irrigation necessitated grander arrangements. Control of the flow of water down the Tigris and Euphrates rivers had to be coordinated by a central authority, so that downstream fields could be watered as well as those further up. It became necessary also to devise a calendar, so as to know when the spring floods might be expected. As these skills evolved, society evolved with them."*

With the expansion of the agricultural metabolism, both in the density of energy and matter flows and in its diversity, society became more complex. Early forms of communal government arose. When farmers grew more food than their families needed, some of them shifted their efforts from the land to the crafts. Barter trade emerged as a new economic paradigm, further stimulating the development of specialised crafts such as weaving, tool making, and pottery [39b]. Some villages grew into cities, which became the centres for trade, government, military, and religion. The first cities probably arose in the plains of southern Mesopotamia as early as the fourth millennium BC [71a]. Subsequently, in a continuous process, the transition from tribal to state societies took shape, as historian Joseph Tainter teaches in his 'The Collapse of Complex Societies' [72a]:

> *"In states, the ruling authority monopolizes sovereignty and delegates all power. The ruling class tends to be professional, and is largely divorced from the bonds of kinship. This ruling class supplies the personnel for government, which is a specialized decision-making organization with a monopoly of force, and with the power to draft war or work, levy and collect taxes, and decree and enforce laws. The government is legitimately constituted, which is to say that a common, society-wide ideology exists that serves in part to validate the political organization of society. And states, of course, are in general larger and more populous than tribal societies, so that social categorization, stratification, and specialization are both possible and necessary.*
>
> *States tend to be overwhelmingly concerned with maintaining their territorial integrity."*

So far as is known, no more than six pristine states appeared: Mesopotamia, Egypt, China, Indus River Valley, Mexico and Peru [72b].

The Agro-Energy Revolution

After that, descendants emerged and a further increase in complexity* among competitors and trade partners arose. Just like in biological evolution, many civilisations rose and fell – *"collapse is a recurrent feature of human societies"* [72c] – such as the first Mesopotamian Empire, the Egyptian Old Kingdom, the Roman Empire, and the Lowland Classic Maya [72d].

In the 15th century a new trajectory of diversification developed, probably catalysed by the invention of the printing press, a device which increased the resources of government enormously:[†]

> *"Laws, for example, could be circulated far more widely and more accurately than ever before. But more important still was the fact that the printing press increased the size of the educated and literate classes. Renaissance civilization thus took a quantum jump, acquiring deeper foundations than any of its predecessors or contemporaries by calling into play the intelligence of more individuals than ever before. But the catch (from a ruler's point of view) was that this development also brought public opinion into being for the first time. Not for much longer would it be enough for kings to win the acquiescence of the nobility and the upper clergy; a new force was at work, as was acknowledged by the frantic attempts of all the monarchies to control and censor the press."*

It was towards the end of 18th century that the dominant form of governing turned 'national', and people were educated in their own mother tongue; not in languages of other civilisations and other times. Societal control became characterised by large, heterogeneous, internally differentiated class structures, a manifestation quite extraordinary in history [72e].

The growing human footprint

Despite the truly remarkable division of labour that began to develop throughout the Agricultural Regime, 95 percent of the people in the world were still peasants [66b]. In the end their hard work did not make life

---

*Tainter understands societal 'complexity' as follows: *"Complexity is generally understood to refer to such things as the size of a society, the number and distinctiveness of its parts, the variety of specialized social roles that it incorporates, the number of distinct social personalities present, and the variety of mechanisms for organizing these into a coherent, functioning whole. Augmenting any of these dimensions increases the complexity of a society. Hunter-gatherer societies (by way of illustrating one contrast in complexity) contain no more than a few dozen distinct social personalities, while modern European censuses recognise 10,000 to 20,000 unique occupational roles, and industrial societies may contain overall more than 1,000,000 different kinds of social personalities."* [72f]

[†]From the Encyclopædia Britannica, under the lemma "government".

healthier or more enjoyable [15i]: the population grew faster than agricultural yields and this brought many people to the edge of starvation [66b]. The societal metabolism continuously needed more food and wood, but also needed more inorganic construction materials. That caused gradual degradation of vital resources of soil, water, grazing land and wildlife. Many productive landscapes converted into barren regions [39b]. Progress took its toll, leaving its tracks on the globe. The severe situation prompted the Reverend Thomas Malthus to publish his famous 'Essay on the Principle of Population' (1798). Malthus argued that human numbers could increase until they were too high for the available food supply. Then famine and disease would reduce the population until it was in balance again with the amount of food that could be produced [66c]. Actually Malthus said, in contemporary human ecology speak, that a population of human beings cannot overshoot the carrying capacity of its habitat.

Genus *Homo* had revolutionarily increased the carrying capacity of the Earth for its own population twice now: through the Pyro-Energy and the Agro-Energy Revolutions. During the Pyrian period, cultural evolution gradually began to overshadow natural evolution in importance, which was a grand revolution in its own right. During the Agrocultural Regime this evolutionary revolution continued forcefully. But before we step into the next period of the evolution of energy regimes, I recall the vital role that fire had continued to play in the history of humankind through the words of Stephen Pyne [49e]:

> *"Fire prepared the fields, fire continually renewed them, fire helped set the rhythms of their cultivation. Fire-floods swamped the native flora and recharged the fields with ashy silt. Fire shocked a site such that, for a while, it could be stocked with exotic wheat, carrots, turnips, cattle, goats, and ragweed. Had farmers shunned fire, the imported cultigens and livestock would have had no advantage over native species. Had they removed fire, the fields might rapidly revert to waste and wild. The ecology of such agriculture was necessarily an ecology of fire."*

Says Goudsblom, *"from the beginning, the process of agrarianization was linked closely to the domestication of fire"* [60b]. To make the first crops cultivated on any large scale easily digestible, *"the farmers needed a hearth to cook on"*. Moreover *"the human predominance over all other mammals"*, partly grounded on the use of fire, formed a precondition for agrarianisation:

> *"The human species' monopoly over fire was so solidly established by the time agriculture began, and is today so easily taken for granted, that it is seldom given separate attention in this context. Yet it deserves mention.*

The Agro-Energy Revolution

> *Their hegemony in the animal kingdom enabled people not only to bring certain species, such as goats and sheep, under direct control, but also – at least as important – to keep most of the remaining 'wild' animals at a distance from their crops and herds."*

And besides the role of fire as a shaping force in land cultivation and thus food production, fire continued to 'shape' its masters [59d]:

> *"Anthropogenic fire demanded a habitation unique to its needs and duties. But humans could not exploit fire without being changed in the process. Fire's abode defined the focus of social life. (In Latin focus means "hearth.") Fire shaped the hearth, and the hearth the house. A family consisted of those who shared a fireside."*

Read 'solar energy release' for 'fire' and it is clear that the sun not only drives the biological, but also the cultural evolution of genus *Homo*.

The domestication of fire holds parallels with the domestication of plants and animals. Whereas the Pyrocultural Regime had been marked by an increasing differentiation between human groups and other animals, the Agricultural Regime showed a continuous differentiation among and within human societies [15i]. And just as with the Pyro-Energy Revolution, the Agro-Energy Revolution gave impetus to social evolution, which in turn boosted the co-evolution of brainpower and technique. However, nature 'paid' for culture through uncountable tree-offerings: during the Agrian very many forests ended up as wood-fuel [15j]. To foster *cultural* development, *Homo sapiens* kept heading for more active and more regular use of *natural* resources, which made him – as an unplanned side effect – continuously more dependent on these very same natural resources. As a result [62g]:

> *"The anthroposphere expanded, and the 'human footprint' could increasingly be spotted by a lunar observer."*

Nevertheless, the impact of the growing agricultural footprint was only a light overture to what was to come in the history of human civilisation, and thus in the evolution of planet Earth.

– 5 –
The Carbo-Energy Revolution

HYDROCARBON MAN

The agricultural energy crisis

Towards the end of the Agrian Period, the steady growth of humankind's societal metabolism fell into stagnation or marginal growth, especially in Europe – *"There was a serious and widespread shortage of wood, which meant that Europe faced an energy crisis"*, says environmental historian Clive Ponting [66d]. Since the origin of fire mastery, wood had largely fuelled [53c] societal metabolism. But the trees could not keep pace with human chopping. By the second half of the 17th century woodlands had become almost a curiosity in certain densely populated regions such as The Netherlands and England [15k]. Ponting reports that [66d]:

> *"By 1717 a newly constructed iron furnace in Wales could not begin production for four years until it had accumulated stocks of charcoal and even then it only had enough fuel to operate for thirty-six weeks before it was forced to close."*

The Dutch and the Flemish found an alternative fuel source in the form of peat to compensate for the lack of wood [15k]. Peat is rather voluminous and therefore relatively costly to carry overland. The Low Countries had an advantageous position with their ubiquitous waterways over which peat could be transported cheaply. The agronomist de Zeeuw has drawn a direct connection between the exploitation of peat and the flourishing period of the Dutch: peat fuelled the societal metabolism that drove their 'golden century'.

Period	
Carbian	About 400 years ago
Agrian	About 12,000 years ago
Pyrian	About 0.5 million years ago
Oxian	About 2.1 billion years ago
Photian	About 3.8 billion years ago
Thermian	About 4.2 billion years ago

The English, not blessed with waterways like their overseas neighbours, adopted another solution [15k]. Since the Middle Ages they had started to replace wood with coal for heating and cooking, and for forging, brewing and making soap [66d]. However, the early coal-innovation induced worrying unplanned side effects [15k]:

> *"One of the first cities to suffer from the 'the gradual elimination of forests at the urban perimeter' was London. As the population doubled from 20,000 in 1200 to 40,000 in 1340, wood as a fuel was increasingly replaced by a low-grade coal called "sea-coal" that was mined near Newcastle and brought to London by ship. Sea-coal emitted a foul-smelling smoke, penetrating everywhere, and left behind an omnipresent deposit of black soot. Its fumes were said to have driven Queen Eleanor out of the city at the time of the feast of Saint Michael in 1257."*

The dawn of ancient sunlight

Following the Black Death in the middle of the 14th century the consumption of sea-coal declined, only to revive again 200 years later when London entered a period of great prosperity and growth [15k]. At that time cardinal innovations of coal-based energy chains appeared. After 1610 new coal clean-up technologies dealt with impurities such that the production of glass and bricks could be undertaken. By 1680 people were beginning to use coke to melt lead, copper, and tin. The next achievement was the industrial production in 1709 of pig iron in a blast furnace and, after 1784, the manufacture of wrought iron. By the end of the 18th century England would have needed many millions of acres of woodland more than the country actually had to be able to supply the wood-equivalent of its annual coal consumption.

The industrial processes just described applied heat released from burning coal to convert raw feedstocks into useful intermediary products. This 'industrial cooking' comprised relatively simple technologies, in effect based on the same principles as domestic cooking, but on a larger scale. The early stages of coal production can hardly be described as mining. *"Many pits were no deeper than 8 to 12 metres."* [73a] When groundwater started to collect at a certain depth, mining stopped at that level. In the 17th century people began to apply water-raising devices, but these machines needed energy to operate. At first water mills, windmills and draught animals were used. But a water mill needs a watercourse to rotate and this was not always available; a windmill needs wind to turn and this does not always blow; and donkeys, oxen, and horses were expensive to maintain. What was needed was an economic and reliable pump. The first technological breakthrough came from Thomas Savery who, in 1702, built a small coal-fired atmospheric steam pump [33j,73a]. Savery's steam pump was the first practical prime mover on Earth to transform chemical energy into motion, or mechanical power. A decade later, in 1712, Thomas Newcomen introduced a five times more powerful steam engine for pumping [33j]. But the first steam engine needed another crucial innovation in 'industrial cooking' [74].

"By 1700 all English trades used coal instead of firewood, with one important exception: iron production." [73b] To understand this, we must look at the double role of coal in iron making: the fuel not only generates heat in processing raw ore; it also acts as a chemical agent in removing oxygen from iron oxides to yield iron (indeed, the oxygen most likely liberated by blue-greens during the Photian). Hence, the coal is both fuel and feedstock. For ages charcoal had been used as the carbon-rich fuel for iron making [33g]. Medieval hearths needed 4 to 8 times more fuel than the mass of ore; that is why the wood demand of medieval and early modern iron smelting led to so many deforested landscapes. Yet, simply substituting coal for charcoal was not a feasible option as the quality of the iron deteriorated due to the chemical impurities in the coal [73b]. Abraham Darby, a British ironmaster, enforced a breakthrough. He produced a high-quality coke from low-sulphur coal. Coke is a product of 'industrial cooking' as it is manufactured by heating coal to a high temperature in the absence of air to drive off volatile constituents, yielding purified carbon. In 1709 Darby killed two birds with one stone. Firstly, the higher strength of coke compared to charcoal enabled more massive ore charges and thus larger furnaces to be used. Secondly, the new process yielded top-notch iron. The quality of Darby's iron made it possible to manufacture thin castings that rivalled brass for pots and other hollowware, such as the cylinder in Thomas Newcomen's steam engine. So, advanced coal-fired cooking in ironworks made production of Newcomen's steam pump possible, which in turn enabled 'deep-shaft mining', thus boosting coal supply. This way the coal-based societal metabolism evolved through industrial symbiosis, or cooperative relationships between several industrial processes.

Newcomen's steam engine did not diffuse successfully into other economic domains, except in some niche domains. The efficiency of the engine was very low: it converted only about half a percent of the coal's energy into mechanical power [33j]. In a coal mine it could run on coal waste, and slowly but steadily – at most, moving at about 12 rotations a minute – keep shafts dry [73a]. An extra complication in finding other applications for Newcomen's steam engine was its reciprocating action; it went up and down, a motion not very suited to driving a factory system based on rotating axles and transmission belts. Indeed, rotating waterwheels and turning windmills powered the dawn of mechanised factory production [15l]. The historian Rolf Peter Sieferle concluded that the mechanical textile industry, with its natural raw materials such as wool and cotton supplied in sailing ships, and its wooden spinning and weaving mills driven primarily by a waterwheel, was *"not so much a pioneer of the industrial system as an outgrowth of the agrarian mode of production"* [73c]. Still, one of the niches through which Newcomen's engine spread outside the coal mines was in the textile sector [73a]: not in driving the factory system, but in augmenting the factory system's prime mover – the waterwheel – by returning the water that had moved the wheel back to a water reservoir, as a pure energy security measure. It is fair to say that during the 18th century, the first century of its existence, the steam engine pumped mainly water.

Towards the end of the 18th century, the rotating steam engine arrived. This was based on James Watt's improved engine design. Although it delivered a relatively small mean power of approximately 20 kW, it surpassed an average contemporary windmill by almost three times, and typical waterwheels by more than five times [33j]. But this lead did not last for too long, as the new invention encouraged competition; by 1830 waterwheels outperformed steam engines once again. It took until the second half of the 19th century before the steam engine broke through [15l,73c]. A range of innovative developments yielded both stationary and mobile applications, from powering belt drives in factories to providing land and waterborne transport in the form of railways and steamships [53d]. The modern steam engine was transportable, adaptable, dependable and durable. Reliable transcontinental networks of transportation came into being [33j] and these, among other things, enabled the import of food. Around 1840 Britain imported about 5 percent of its food, but by the end of the century 80 percent of the grain for human consumption was imported, together with 40 percent of meat, and 70 percent of dairy products. This foreign food supply provided a solution to the perennial problem of how to feed a rising population [66e]. Coal-fired steamships began to beat the till then ever-improving sailing ships [33k]; the bulkiness of steam engines was much less of a problem on water than on land – *"During the late 1830s steamships could not beat sailships in a brisk wind. A decade later the best steamships cut the translantic crossing to less than ten days."* Steamships made large areas of America and Australia accessible for Europeans. *"Only then did migration on a scale worth mentioning occur"* [73d] and America began to develop into a laboratory for innovative industrialisation. Later coal-fired steel production enabled larger ships to be built [33k]. But for services across short distances steam machines did *"not displace the horse by any means"*. In Germany, for example, the number of horses rose from 2.7 million at the beginning of the 19th century to 4.6 million in 1913 [73c]. *"Only the massive spread of the automobile and the tractor during the 20th century finally pushed the horse into the leisure sector."*

The steam engine's heyday was limited to the second half of the 19th century [15l]. Just when it had beaten waterwheels, the next generation of heat converters attained superiority [53d]. For more than a century contemporary historians viewed the steam engine as *"the pivot on which industry swung into the modern age"*, describing its invention as *"the central fact in the industrial revolution"* [15l]. Yet today's historians conclude their predecessors have overestimated the role of the steam engine. In the year 1800 Watt-machines consumed 3.25 percent of the British coal production [73c]. That figure grew substantially in the first half of the 19th century, largely as a result of the growing *"steam engine complex"* as a whole, that is steam engine-driven mobility together with ironworks for producing the movers (locomotives and boats) and the infrastructure (railways and bridges). It was the output of ironworks defining *"the mechanical civilisation of nineteenth-century Europe"*, says Vaclav Smil [53e]. The crux was coal-fired industrialisation [73c]:

> *"Considering the central importance of coal as the energy basis of the Industrial Revolution, it is quite astonishing that ... it has been almost completely ignored by economic history."*

Or, differently formulated [73e]:

> *"Without coal, European societies of the 18th and 19th centuries would have remained agrarian societies, even if they had utilised the innovation potential fundamentally embodied in agrarian societies to a much greater extent."*

In terms of evolutionary energetics, coal firing fiercely boosted *Homo sapiens'* societal metabolism. An unparalleled increase in diversity and complexity of anthropogenic coal-based energy chains arose. Nowadays, the transition of the energy economy from wood to coal is seen as *the* driving force of the late 18th century Industrial Revolution [15m,33l,39b,59e,66f,75a], defined as the *"process of change from an agrarian, handicraft economy to one dominated by industry and machine manufacture"* [76]. But coal-fired industrial cooking arose before, *and enabled*, coal-fired industrial motion. From then on the industrial symbiosis of mechanised factory systems – as developed in the 'agrarian' textile sector – with coal-fired power accelerated industrial machine manufacturing. The Industrial Revolution emerged on the wings of a coal-fired energy revolution.

The Carbocultural Regime

Coal use stimulated a rapid transformation of England's energy economy, a truly historic transition in terms of energetics as the English chemist and Nobel laureate Frederick Soddy (1877 to 1956) realised [53f]:

> *"Wind power, water power, and wood fuel are parts of the year-to-year revenue of sunshine ... But when coal became king, the sunlight of a hundred million years added itself to that of today and by it was built a civilization such as the world had never seen."*

Coals consist of sedimentary rocks formed largely by lithification of peat, which is itself produced from the accumulations of dead plant matter in wetlands [33m,53g]. The black solids store solar energy harvested by green plants hundreds of millions years ago.* Viewing coal as 'fossilised ancient sunlight' makes it clear that coal is not a primary source of energy, but a *re*-source. The sun is the true primary source, because a source flows whereas a *re*-source conserves. The winning of coal for industrial combustion applications heralded a solar energy revolution, the Carbo-Energy Revolution; 'carbo' stands for carbon, the chemical element that characterises fossil fuels such as coal, natural gas, crude and shale oil, and tar sands, all so-called hydrocarbons.

For the fifth successive time during the evolution of energy regimes, life revolutionarily increased the utilisation of solar energy for its own development by at least an order of magnitude. A new energy period began, the Carbian Period, during which a new societal metabolism, or energy regime established, the Carbocultural Regime. The Carbian began around the year 1600 with anthropogenic coal mining and combustion, which drove the emerging metabolism, especially the processing of raw materials through 'industrial cooking'. This dating is consistent with the evolution of coal production [53e]:[†]

---

*Fossil fuels are formed, as a result of geologic processes, from the remains of organic matter produced by photosynthesis hundreds of millions of years earlier. The lithification of peat-yielding coal is such a geologic process. But fossilisation processes extract subterrestrial heat to run. Therefore, fossil fuels store geothermal energy too – maybe just a few percent – in addition to solar energy.

[†]The beginning of the Carbian Period does not mark the first use of coal. *"The Romans already used coal, probably for forging metals"* [73f] and there is evidence that coal was not unknown to the Greeks. The Chinese also used coal at an early stage, no later than the 13th century, but *"no commercial or industrial dynamic unfolded in China"*. England's first coal use in London, during the Middle Ages, played no role in quantitative terms and was domestic (heating and cooking) or small-scale (forging) by nature. The Carbian Period began with industrial coal use.

The Carbo-Energy Revolution

"Between 1540 and 1640 almost all English coalfields were opened up for exploitation; by 1650 the country's annual coal output passed 2 Mt, 3 Mt annually were extracted in the early eighteenth century, and over 10 Mt were extracted annually by its end."

After the Pyrian and the Agrian, the Carbian is the third energy period of the Pyroic, wherein the third wave of fire mastery emerged. During the Carbocultural Regime *Homo sapiens* reinforced ecological dominance, or more precisely formulated, the cultural – not biological – subspecies *Homo sapiens carbonius* did [77a]. Like Man the Fire Master is an icon for human's first ecological dominancy and divergence from all animals, and Man the Solar Farmer for *Homo sapiens'* capacity to cultivate land, Hydrocarbon Man symbolises the unequalled anthropogenic force extracted from fossilised, ancient biomass.

THE CARBIAN EXPLOSION

An explosion of competing energy chains

Although contemporary economic historians see the steam engine as just *an important development in the launching of the Industrial Revolution*, for an energeticist its invention still marks an earth-shaking event: it was the first device enabling heat to be converted into mechanical work. Just imagine, societies in those days suffered from a severe shortage of power. Despite earlier innovations in wind and water power supply, the main source was still the human 'biological engine', or muscle – not an animal's, as food shortages constrained its use to only badly needed applications [66f]. The early industrial applications of heat from burning coal, as used in glass making and ironworks, involved converting heat into chemical work, not mechanical work. Metabolically speaking, biological species like animals and human beings produce *bio*-chemical and *bio*-mechanical work through the combustion of fresh carbohydrates in their cells, while carbocultural communities produce '*anthropo*-chemical' and '*anthropo*-mechanical' work through the combustion of fossil hydrocarbons in their societies.

The great idea of the steam engine – to create motion by converting heat into power – provided a tremendous stimulus to innovators. Visionaries saw a world full of combustion engines energised by the rich deposits of the fossil combustibles [33j]. This would free man's arm, literally. And indeed, the next generation of heat converters, such as the four-stroke internal combustion engine and the steam turbine, were better than steam engines [53d]. The former provided convenient and compact power for road transport, and the latter efficient generation of electricity.

Electricity represents a metabolic revolution unlike any other in the genesis of man-made energy chains [53h]. A whole new infrastructure had to be put in place before a break-through idea could become viable: the generation of electrical current, its distribution to end-users, and the provision of electricity-powered appliances. Society accepted the new, non-fuel energy carrier almost immediately: it was clean, silent and odourless at the plug – thus convenient. The first two electricity-powered devices, the electric lamp and the electric motor, almost immediately modernised industrial processes and human lifestyle – and radically, too.

Before the electric lamp came on the market, lamps burned kerosene (made from oil), or town gas (made from coal) to meet the demand for artificial light. Previously, oil lamps had burned whale oil, a biofuel. Mineral oil had been known for millennia, for example in the Middle East from natural seepages and pools. But some considered it a nuisance, others a miracle cure [78], and again others used it as a protective coating [33n]. It was not burned to extract the stored solar energy.* But things changed drastically after 1849, the year in which Abraham Gesner invented a technology to distil kerosene from petroleum [78]:

"Gesner's kerosene was cheap, easy to produce, could be burned in existing lamps, and did not produce an offensive odor as did most whale oil. It could be stored indefinitely, unlike whale oil, which would eventually spoil. The American petroleum boom began in the 1850s. By the end of the decade there were thirty kerosene plants operating in the United States. The cheaper, more efficient fuel began to drive whale oil out of the market."

Gesner's invention enabled people to convert ancient sunlight into artificial light. The consequential rising demand for petroleum inspired Colonel Edwin Drake to explore the subsurface for this resource. He began drilling in 1858 and struck oil a year later [33n]. His drilling techniques spread quickly; Titusville, the scene of the colonel's activity, and other northwestern Pennsylvania communities became boom towns [79]. Vaclav Smil notes that [33n]:

"By 1900 two other American states (Texas and California), Romania (the Ploesti fields), Russia (the huge Baku fields on the Caspian Sea), and Dutch Indonesia (Sumatra) led the ranks of crude oil producers. During

---

"Although there is a good case for abiogenic origin of some hydrocarbons, most fossil fuels are clearly organic mineraloids with minor quantities of inorganic contaminants." [53g] Most fossil fuels, which appear in solid, liquid or gaseous forms, are biogenic by nature, meaning that they derive from ancient, fossilised photosynthetic organisms. Hence fossil fuels are stores of 'ancient sunlight'.

The Carbo-Energy Revolution

the first two decades of the twentieth century they were joined by Mexico, Iran, Trinidad, and Venezuela (1914)."

With the increase in availability of kerosene, the whale oil market collapsed: the whaling fleet of 735 ships in 1846 had shrunk to 39 by 1876 [78]. James Robbins' conclusion that *"capitalism saved the whales"* is as true as libertine; saving the whales was an unplanned side effect of innovating the oil-to-light energy chain, which continued to evolve rapidly:

"A new invention soon snuffed out both flame-based systems. In 1879 Thomas A. Edison began marketing the incandescent light bulb he had invented the previous year. Arc-light technologies had existed since the turn of the century, but it was Edison who devised the modern, commercially feasible light bulb, which produced an even light, burned longer and brighter than oil or kerosene, and was much safer than an open flame. As the country was electrified, whale oil and kerosene were both driven from the illumination market."

But instead of dying a gentle death together with the oil lamp, the oil industry adapted itself to the rapid diffusion of internal combustion engines in cars, ships, planes, and agricultural machinery. Gasoline engines induced the emergence of the largest manufacturing industry of the 20th century, car making. The degree of personal mobility the car offered appeared an unbeatable value proposition from its very beginning. Diesel engines proved invaluable in trains, in shipping, and in heavy vehicles [33k]. Gas turbines, invented in the 1930s, were adopted by aeroplane manufacturers, which led to the advent of mass international flying.

The internal combustion engine became the prime mover for mobility applications, the steam turbine for stationary power, and the electric motor revolutionised manufacturing [33s]:

"The distribution of steam-generated mechanical power for processing and machining did not differ from the pattern established by waterwheels: gears rotated long iron and steel line shafts passing along factory ceilings, and pulleys and leather belts transmitted this motion to parallel countershafts belted to individual machines, often also to different floors through holes in the ceiling. Any mishap – prime-mover break down, damaged gearing, cracked shafts, slipped belts – stopped the whole assembly."

Electric motors began to dominate in factories after 1900. These devices, together with electric lighting, brought enormous gains in productivity as well as improvements in the consistency and quality of finished products.

Then another branch of electricity applications arose. The invention of the telegraph and telephone, both dating back to the 19th century, enabled instant remote communication by wire [33o]. The first American transcontinental telephone link came in 1915, while the first transatlantic telephone cable was laid in 1956. Until about 1930 Samuel Morse's famous codes dominated intercontinental, wireless tele-messaging. Thereafter, the radio began to replace the radiotelegraph, introducing the next innovation of fossil fuel-based communication. A transmitter converted electrical energy into radio signals and a receiver converted the radio signals 'back' into sound, or radiative energy into sonic energy. Radio sound emerged from a coal-based energy chain in the form of a socio-metabolic process.

In addition to the innovations in manufacturing, transport, and communication technology described above, humankind also revolutionised its food economy again, this time by augmenting crop cultivation and animal breeding with fossil-fuel-based innovations. *Homo sapiens carbonius* began to boost his solar farming activities with 'subsidies' of ancient sunlight, for example through the production of chemical fertilisers [53i]. Besides adequate water supply, nutrients are the critical inputs opening photosynthetic work gates. Nitrogen is very important and although this element dominates the atmosphere, its presence in the biosphere is relatively small. In 1913, the manufacture of nitrogen fertilisers became possible with the Born-Haber process for synthesising ammonia. Nonetheless the onset of large-scale fertilisation did not occur until the early 1950s, when new, artificially selected plant species could take full advantage of intensive fertilisation. Since then, the use of the three main nutrients – nitrogen, phosphorus and potassium – has grown exponentially.

In 1900, fossil energy subsidies in agriculture were practically negligible; in 1985 about one-third of the energy contained by the food that *Homo sapiens carbonius* consumed was fossil; thus about one-third of the biochemical and biomechanical work that his body expended, from the pounding of his heart, to the words he spoke, and the hand he moved, was derived from fossil energy – of course, today that will be not much different. Without these energy subsidies a slowly rising cultivated area perhaps could have supported a global population of about 2.5 billion human beings in 1985 [53j]. But the actual number was 4.8 billion people [53k]. 'Carbo-agriculture' increased food production by almost a factor of 6 within 85 years! The corresponding population growth, albeit 'only' with a factor of 2.8, was the most drastic in the history of humankind.

Electric energy as information carrier

A new family of electric energy converters came into existence with the invention of the digital computer during the 1940s when the *"exigencies of war gave impetus and funding to computer research"* [80]. In Britain, the incentive was to crack ciphers and codes produced by German electromechanical devices such as the Enigma machine and the *Geheimscheiber* (Secret Writer). In the United States, work began in early 1943 on the Electronic Numerical Integrator and Computer (ENIAC) aimed at generating artillery range tables. But when delivered in 1946, *"the war it was designed to help win was over"*; ENIAC's first task became to perform *"calculations for the construction of a hydrogen bomb"*. The ENIAC was an enormous machine, having a physical footprint of 9 by 15 metres; *"with approximately 18,000 vacuum tubes, 70,000 resistors, 10,000 capacitors, 6,000 switches, and 1500 relays, it was easily the most complex electronic system theretofore built"*. And it performed at a speed that was at that time beyond human ability.

Paul Freiberger and Michael Swaine [80] discriminate three movements in the surging digital computing wave that followed: the rise of mainframes causing the first movement, of 'Many persons, one computer'. Personal computers unleashed the second movement, of 'One person, one computer', which overshadowed the first soon. The third movement, still ongoing, is about embedding microprocessors. Today the general-purpose microprocessor contributes only about 20 percent of the annual sales in the semiconductor business, whereas special-purpose processors, controllers, and digital signal processors have shown the greatest growth.

> *"These computer chips are increasingly being included, or embedded, in a vast array of consumer devices, including pagers, mobile telephones, automobiles, televisions, digital cameras, kitchen appliances, video games, and toys."*

Digital computers are designed to be connected, and connected computers form a network. The first Internet, a network of remote computers, became operational in 1969 in the United States and connected 15 universities. Later Tim Berners-Lee and others developed a communication protocol enabling the creation of the so-called world-wide web, in 1991, which grew exponentially almost from its incubation.

> *"Originally created as a closed network for researchers, the Internet was suddenly a new public medium for information. It became the home of virtual shopping malls, bookstores, stockbrokers, newspapers, and entertainment. Schools were "getting connected" to the Internet, and children were learning to do research in novel ways. The combination of the Internet, E-mail, and small and affordable computing and communication devices began to change many aspects of society."*

The world-wide web has expanded swiftly: the number of 'host' computers grew from 376,000 in 1990 to 72,398,000 in 1999 [81a]. Apart from personal computers, devices with embedded microprocessors can be connected too, even globally; think, for instance, of the Global Positioning System (GPS). Some speak in this respect about ubiquitous computing, a phenomenon which already extends the increasingly networked world.

In terms of evolutionary energetics, computers and their networks are extraordinary realisations. All the devices connected are energy converters, but they do not convert a natural flow of energy into an energy resource, like a green plant, or convert heat into motion, like an engine, nor do they convert motion into electricity, like a dynamo. Microprocessors convert normal electric current into encoded electric current. Thus electricity, as an energy carrier, becomes an information carrier as well.* In effect, digital energy converters add information to an electric current by modulating it.

Mobile digital devices tap electricity from batteries. And even though the performance of batteries has steadily improved, mobile electronics evolve at such a speed that the small chemical power plants struggle to keep up. *"Batteries have become the showstopper in today's wireless world ... They don't provide enough power, they're too heavy for the power they do provide, and they don't last long enough to meet the demands of the next generation of portable digital devices."* [82] Today lithium-ion batteries are the systems of choice, because of a their relatively high-energy density, flexible and lightweight design, and longer lifespan than comparable battery technologies [83]. They are 'rechargeables', charged again and again with electricity from a plug remotely connected to a power plant – of which the majority burn fossil fuels. The fascinating virtual world thrives largely on ancient sunlight.

Carbocultural non-fossil energy chains

The Carbian Explosion of diversity in fossil-fuelled energy chains also spun off a few viable non-fossil systems. The oldest one is hydroelectric power. In 1833, Benoit Fourneyron radically improved the waterwheel by inventing the water turbine [33p]. In 1849, James B. Francis patented the spiral, inward-flow turbine, which has become by far the most important hydraulic mover in the industrial age. *"First water turbines powered directly industrial machinery, most notably textile mills"*, but the breakthrough came *"once they were coupled with electricity generators"*. Hydroelectricity is the result of industrial symbiosis between a descendant of the waterwheel,

---

*This is not to say that the microprocessor was the first device encoding electric current – telegraph and telephone preceded, for example – but, as a miniature electronic device, it enabled the development of microcomputing.

The Carbo-Energy Revolution

the water turbine, and a power generator. In 2001, hydropower plants all over the world together contributed about one-sixth of the global electricity supply [84a].

The second significant, non-fossil energy chain is nuclear power. In nuclear reactors, nuclear fission occurs: a relatively heavy nucleus of an atom breaks up into lighter nuclei, a process in which a large quantity of energy is released, radioactive products are formed, and several neutrons are emitted [85]. As nuclei rarely split spontaneously, their splitting must be induced. Otto Hahn and Fritz Strassmann discovered in 1938 that the uranium atom split when bombarded by 'slow' neutrons [86]. Almost immediately (early 1939) Enrico Fermi suggested that neutrons might be among the fission products of uranium fission, so raising the image of a chain reaction in the minds of scientists – *"If controlled in a nuclear reactor, such a chain reaction can provide power for society's benefit. If uncontrolled, as in the case of the so-called atomic bomb, it can lead to an explosion of awesome destructive force."* In 1942, on December 2, the first human-induced sustained nuclear chain reaction on Earth came into being [33p]. Some 14 years later, in the United States the first commercial nuclear power plant began generating electricity and highly radioactive waste (which is not processed, but stored instead). In 2001, nuclear fission contributed 17.1 percent to worldwide power production [84b].

The third non-fossil energy chain is one directly driven by energy from the sun, as explored already by the Romans. Historic remnants seem to reveal that they trapped solar heat, in *heliocamini*, or 'sun furnaces', by exposing a window of clear glass in a closed room, faced south, to the sun [87a]. But window glass, and thus the sun furnace, did not survive the fall of Rome. In 1767, Horace de Sauserre built a miniature greenhouse that brought water to the boil. This invention in turn inspired, a century later, Augustin Mouchot who wanted to produce electricity from solar energy. One of his sun machines, a solar reflector designed to collect heat, drove a steam machine at the Universal Exposition in Paris in 1878. However, a French government commission ruled after a year of testing that *"In France, as well as in other temperate regions, the amount of solar radiation is too weak for us to hope to apply it … for industrial purposes."*

During the late 1870s, two British scientists, Adams and Day, observed that sunlight could cause *"a flow of electricity"* in selenium bars [87b]. A few years later the renowned German scientist Werner von Siemens presented selenium modules, prepared by Charles Fritts, to the Royal Acadamy of Prussia. However, the large majority viewed Fritts' 'magic' selenium plates *"as perpetual motion machines"* and simply dismissed the so-called 'photoelectric effect'. Until, in 1905, Albert Einstein prepared the human mind for its scientific understanding:

> *"[He] showed that light possesses an attribute that earlier scientists had not recognized. Light, Einstein discovered, contains packets of energy, which he called light quanta (and which we refer to as photons)."*

By the 1920s scientists saw that certain photons could kick certain electrons from a semiconductor to the so-called conduction band, which allowed them to be set into motion by a voltage, yielding an electric current [88]. In spring 1953, two Bell scientists led the pioneering effort *"that took the silicon transistor ... from theory to working device"* [87c]. One of them, Gerald Pearson, whilst experimenting *"inadvertently made a silicon solar cell"*, which was far more efficient than the selenium solar cell. However, the high price of silicon blocked the road to commercial applications [87d]. But profit and loss have a different meaning to the army than in business, and hence the notion 'cost' does. The U.S. Army Signal Corps thought of an application, top-secretly dubbed 'Operation Lunch Box': the construction and launching of a communication satellite [87e]. In 1955 President Eisenhower announced America's plans to put a satellite into space, and for such a vehicle solar energy was the logical power source. In 1958 the satellite went into orbit, and the solar cells did what they where expected to do: powering the transistor radio on board [87f]

Photovoltaics do not exploit the sun's heat, but its light [87g]:

> *"[It is] the direct conversion of the sun's energy into electricity ... Hardly the rugged stuff utility people are accustomed to, photovoltaics does away with the bulky paraphernalia – boilers, turbines, pipes, and cooling towers – required by all other electricity-generating technologies. In fact, solar cells operate without moving parts. Within those few microns, photons, packets of energy from the sun, silently push electrons out of the cells and so make electricity."*

Photovoltaics, in its present-day shape, is a truly carbocultural phenomenon. The basic component, silicon, does not occur uncombined in nature. It is found in practically all rocks as well as in sand, clays, and soils, combined either with oxygen as silica (silicon dioxide), or with oxygen and other elements (e.g., aluminium, magnesium, calcium, sodium, potassium, or iron) as silicates [89]. Like ironworks, 'silicon-works' implies the removal of oxygen from raw oxides, silicon oxides in this case. Thus, just as hydro- and nuclear power, and the steam engine centuries ago, solar power needs 'industrial cooking' to come into existence for preparing the artefact's construction ingredients. And 'industrial cooking' needs 'industrial fire' from fossil fuels releasing ancient sunlight in the form of heat. When *Homo sapiens carbonius* wishes to harness incident sunlight he begins to exploit an ancient form. Today's dominating socio-metabolism is carbocultural at its core.

The Carbo-Energy Revolution

Sensitive scientific senses

Both during the Agrian Period and the Carbian Period increased energy utilisation went hand in hand with population growth. However, this observation does not permit the conclusion that a higher energy utilisation *causes* population growth; it is a precondition, indeed, but not the single determining factor – all evolution is co-evolution. For instance, in the previous chapter we have seen that farming brought higher food – energy – yields than conventional hunting and gathering, so stimulating population growth, but in addition, reasons Jared Diamond, the sedentary lifestyle of early farmers allowed women to bear and raise as many children as they could feed, while hunter-gatherer mothers were constrained by problems of carrying young children on treks [90a]. Likewise, during the Carbian Period the phenomenal fossil energy subsidies for food production pushed up population densities, but better sanitation and health care also increased the life expectancies of people. *Homo sapiens carbonius* began to add scientific knowledge to his living.

According to Edward O. Wilson [69b]:

> *"Today the greatest divide within humanity is not between races, or religions, or even, as widely believed, between the literate and the illiterate. It is the chasm that separates scientific from prescientific cultures."*

A world without scientific knowledge would be a world without fertilisers, wireless communication and pacemakers, and without supermarkets filled with food. A world without scientific knowledge added to human artefacts is simply uninhabitable to *Homo sapiens carbonius*.

The way scientists work provides a unique characteristic of the Carbian Period. Scientists, in close cooperation with technologists, persistently expand their senses with a special kind of artefact, the measuring device. *"Without the instruments and accumulated knowledge of the natural sciences – physics, chemistry, and biology – humans are trapped in a cognitive prison"*, contends Wilson; *"once we were blind; now we can see – literally"*, and *"once we were deaf; now we can hear everything"* [69b]. The human eye senses light with wavelengths between 400 and 700 nano-metres (billionths of a metre), but *"scientists have entered the world of animals and beyond ... The largest galactic cluster is larger than the smallest known particle by a factor of the number one with about thirty-seven zeroes following it."* The human ear senses in the range of 20 to 20,000 Hz, but *"now, with receivers, transformers, and night-time photography* [scientists] *can follow every squeak and aerial roll-out."*

The *"search for the ultimate"* has been advancing since the origin of microscopy in the late 1600s [69c]:

> *"This technological enterprise satisfies [an] elemental craving: to see all the world with our own eyes. The most powerful of modern instruments, invented during the 1980s, are the scanning-tunneling microscope and the atomic force microscope, which provide an almost literal view of atoms bonded into molecules. A DNA double helix can now be viewed exactly as it is."*

The passion for dissecting and reassembling, Wilson argues, has resulted in the invention of nano-technology, an all-embracing concept referring to the capacity of human beings to manipulate, modify or mould matter on the nano-scale. Besides nano-technology, nowadays biotechnology, understood as biomolecular engineering, and computer and communication technology are often crowned with the label 'revolutionary' [91, 92]. However, it is more insightful to consider them as two incarnations of the nano-technological advancement. Genetic engineering, for example, boils down to the molecular modification of DNA material while the rapid miniaturisation, during the past decades, of semiconductors brought microprocessor manufacturing into the 'nano-transistor' regime [93]; thus transistor dimensions now approach the dimensions of biological molecules.

The nano-technological achievement builds on fundamental knowledge about physical, chemical, and biological processes happening at the nano-scale, in turn, enabled by the capacity *"to see all the world with our own eyes"* through self-made instruments. Again, enhancing the sensory experience by instrumentation is a precondition to realising 'nano-works', but is not the only determining factor. The evolution of science is a co-evolutionary act between 'observing' and 'theorising': theory makes sense of what the senses and their artefactual extensions observe [69c]. In the previous chapter we have seen that *"the cutting edge of science is reductionism, the breaking apart of nature into its natural constituents"*; according to Wilson *"the explanations of different phenomena most likely to survive are those that can be connected and proved consistent with one another"*, a phenomenon he calls *"consilience"*. For example:

> *"Quantum physics thus blends into chemical physics, which explains atomic bonding and chemical reactions, which form the foundation of molecular biology, which demystifies cell biology."*

Carboculture stirred up the demystification of natural phenomena by science and thus the anthropogenic, artefactual (r)evolution. The apotheosis seems nano-control, with genetically modified organisms, shape-selective catalysts and fast microprocessors as expressions of it. In the Carbian today, the scientific perception of the world is probably farther away from daily experience than ever before, but, at the same time, the achievements of science have never been so pervasive and qualifying in modern life.

The Carbo-Energy Revolution

AND THE FACE OF THE EARTH CHANGED

The surging human footprint

The face of the Earth has changed radically during the Carbian Period. Vaclav Smil writes [33l]:

> *"The single word capturing best the essence of modern societies is growth: growth of energy output leading to growth of cities, populations, crop yields, industries, economies, affluence, travel, information, armaments, war casualties, and environmental pollution."*

During the Carbian, *Homo sapiens carbonius* revolutionarily increased the carrying capacity of the Earth for its own population. In 1825, the human population reached the 1 billion mark [94a]; in 2000 this number was 6.1 billion [95a]. The global socio-metabolism, which is the flow of energy and matter through society at large, increased phenomenally and unprecedentedly. The total world energy consumption grew almost 14-fold between 1900 and 2000. *"In many ways, the defining economic development of this century is the harnessing of the energy in fossil fuels"* [94a], which today dominates the energy economy, accounting for 86 percent of primary commercial energy markets [95a]. We have already seen that fossil-fuelled agriculture, augmented in particular with scientific knowledge and modern technology, yielded an increase in food production by about a factor of 6 in the 20th century [53j].

Material use became increasingly more complex: today's stock, for example, draws from all 90 naturally occurring chemical elements in the periodic table, compared with just 20 or so at the beginning of the 20th century [94b]. But perhaps the rise in volume is even more impressive. In 1995, nearly 10 billion tonnes of materials – industrial and construction minerals, metals, wood products, and synthetic materials – entered the global economy. This is about two-and-a-half times more than in the early 1960s, when global data were available for all major categories. In the U.S. materials consumption has grown 18-fold in the last century; around the turn of the millenium, an average American 'used' 37 tonnes of material.

> *"Humans worldwide move 40 billion tons of earth each year – churning it up during mining operations and construction and indirectly eroding it away by ploughing croplands and clearcutting forests ... Only rivers, surging and meandering through their flood-plains, constantly remodelling their channels and banks, routinely rival humans with bulldozers for tonnage of earth moved."* [97a]

Even with surging recycling activity, most materials moving through industrial economies are used only once, and are then thrown away. The Worldwatch Institute says provocatively [94b]:

"An extraterrestrial observer might conclude that conversion of raw materials to wastes is the real purpose of human economic activity."

Apart from moving matter, *Homo sapiens carbonius* boosted its own mobility. In the early 19th century, when people still lived on farms, *"travel was not much different than it had been 1,000 years earlier, limited to the speed of a horse"* [94a], but these days Hydrocarbon Man *"routinely orbit the Earth in 90 minutes"*. At the end of the 19th century, early steam-powered trains and the first motorcars with internal combustion engines went no faster than 40 kilometres per hour – *"and their high cost kept most people on foot"*. But during the past century the number of automobiles has grown from a few thousand to half a billion, and aviation has developed from the pioneering flight by the Wright brothers in 1903 to jet aircraft flying faster than sound, moving 400 passengers at once over the ocean.

The true power of the Carbian Explosion comes out most powerfully in a historic perspective:

"The growth in economic output in just three years – from 1995 to 1998 – exceeded that during the 10,000 years from the beginning of agriculture until 1900. And growth of the global economy in 1997 alone easily exceeded that during the 17th century."

Advancements in electronic information exchange added new dimensions to the process of globalisation. The number of telephone lines has increased more than 8-fold since 1960, the number of cell phone subscribers rose 13-fold between 1990 and 1996 and *"at the end of 1998, the world's first affordable satellite telephones went on the market, bringing the world's most remote regions into the ubiquitous information web"*. Booming world travel and global communication facilitated the rise of supranational and worldwide organisations, such as the United Nations with its many specialised agencies such as the World Bank, and countless others. The number of non-governmental organisations, together civil society, also grew considerably, from virtually none around 1950 to over 40,000 half a century later [98a].

Although the first cities had arisen millennia earlier, in 1850 only two percent of the people lived in a city [39c]. During the Carbian Period, farming began to free rural labour forces to the growing manufacturing sector, partly as a result of the fossil energy subsidies. A rapid urbanisation started. In 1900, 16 cities counted a million people or more, and roughly 10 percent of humanity dwelled in cities [94a]. Today, there are over 300 multimillion cities and 14 mega-cities housing more than 10 million residents. In 1995,

The Carbo-Energy Revolution

75 percent of the human population in industrialised countries lived in urban areas; for the world population as a whole, this amounted to 45 percent [94c]. A satellite picture taken today at night shows a myriad of concentrated light spots, areas of intensive energy dissipation reflecting omnipresent city life. Humanity stands on the verge of an historic transition from living predominantly in a rural setting to a predominantly urban one. This 'dwelling revolution' not only changed the world from landscape to cityscape, it also changed the experience of the world and the 'metabolic awareness' of its citizens. Whereas Man the Solar Farmer produced 'natural' food and wood himself, Hydrocarbon Man buys processed food and fuel in the city.

The accelerating growth of the anthroposphere, or the Carbian Explosion, has become a global force. During the Pyrian Period genus *Homo* continuously expanded its ecological domain. Environmental changes in the human habitat have grown from local to regional and recently to global phenomena, as issues such as ozone depletion and global warming illustrate. Likewise, *Homo sapiens'* interference with the life of other species has spread globally. There is little doubt that the Earth's biodiversity is declining [95a,99]. It is estimated that – today – species are disappearing 100 to 1000 times faster than they did before the arrival of humans, and according to the biologist Stuart Pimm this rate is going to accelerate in the coming years [28u]. The Living Planet Index* fell by 37 percent between 1970 and 2000 [96b]. In addition, the so-called ecological footprint, a measure of man's use of renewable natural resources, grew by 80 percent between 1961 and 1999 to an estimated level of 30 percent above the Earth's carrying capacity [95a,96b].

From fire master to fire addict

Fire Mastery made genus *Homo* ecologically dominant. First *Homo heidelbergensis* and *erectus* domesticated 'wild' fire. Then *Homo sapiens carbonius* encapsulated fire by industrialising it. However, in evolving from Fire Master to Hydrocarbon Man, genus *Homo* became more and more addicted to the heat released by burning fuels [58d]:

> *"Remove fire from a society, even today, and both its technology and its social order will lie in ruins."*

---

*The World Wide Fund For Nature (WWF) coordinates a partnership of scientists keeping track of the so-called Living Planet Index, which is the average of three sub-indices measuring changes in forest, freshwater, and marine ecosystems [96].

The sociologist Johan Goudsblom labels us, human beings, metaphorically as 'pyrophytes', or 'fire-growers'; without fire we would be either overgrown by competitors or unable to reproduce ourselves [15n]. He observes that [15o]:

"The trends … towards increasing use of fire, in a more concentrated form, under conditions of continuously advancing specialization and organization have contrived to make the control of fire seemingly more simple but actually far more complex. As a result of these trends, more and larger fires have been caused by humans in the twentieth century than in any previous age. People have derived more comfort from fire than ever before, and they have inflicted greater damage and suffering with it."

Fire mastery unleashed seemingly magical forces in only a flash of the history of life on planet Earth. However, while Hydrocarbon Man – like none of his predecessors – may well control every single fire with a clear view of the target in mind, he does not control the inherently unplanned side effects of all his fires together. Industrial fire has developed its own momentum, it seems, and may surprise the master through causing repercussions. The biologist David Burney portrays today's state of the world, at today's juncture in the Pyroic, as follows [28v]:

"Evolution has now entered a new mode. Something altogether new is happening, and it has to do with what humans do to the evolutionary process. And it's a very scary thing, because it's like we are taking evolution around a blind corner, something that nature hasn't dealt with before: species that can just hop a plane and wind up on the other side of the world; combinations of species that have never been combined before. It's a whole new ball game, and we don't know, really, where it will end."

At this point in the story, we have reached this very day, and thus the end of our time travel through the history of life along the tracks left by energy. Subspecies *Homo sapiens carbonius* dominates planet Earth ecologically. And his global force, derived from ancient sunlight, is still developing powerfully. Right now all living species are experiencing the Carbian Explosion, one way or another. *Homo sapiens* could already be overshooting the carrying capacity of the Earth for his own population, or he will do so tomorrow when the course of his societal development remains 'business as usual'. Indeed, the human advantage over the animal kingdom, fire mastery, brought genus *Homo* evolutionary dominance in a record period of time, but also a footprint (soon to be) too large to be carried by planet Earth – that is why current carbocultural development is ecologically not sustainable, a notion putting forward a compelling case for change.

PART II
THE NEXT ENERGY REVOLUTION
— evolutionary energetics, models and scenarios —

– 6 –
The Staircase of Energy Regimes

ENERGY REGIMES

The foundation of the Staircase model

The Staircase of Energy Regimes is a conceptual model based on the relationships between energy, complexity and evolution. (These relationships are described in more detail in the Appendix, while the text box 'Modelling the future' discusses some aspects of conceptual models *versus* computer models.) The Staircase aims to model the evolution of life on planet Earth from an energy perspective. Therefore, to begin with, we need to have some understanding for the nature of energy itself.

Energy is *"the capacity to do work"* [100a], or the capacity to cause change. For *change* to occur, energy must transform into another quality. The Second Law of Thermodynamics, that is *"the study of the transformations of energy in all its forms"*, holds that natural, and thus spontaneous, change accompanies the fall of energy from a higher to a lower quality; energy does not vanish, but is conserved, as the First Law states. The nature of energy is to decay, to dissipate, to produce 'chaos', or entropy. However, having the *tendency to dissipate* is one thing, *being able to dissipate* is quite another. Energy 'wants' to produce entropy, but it 'needs' *dissipative paths* in the physical world to do so.

Low-entropy sunlight trapped by the Earth's atmosphere seeks dissipation paths to escape again into the cold outer space in the form of relatively high-entropy radiant heat; without such a flow the planet would continuously heat up. A tiny, tiny, bit of incident sunlight drives the photosynthetic processes on Earth. At first sight, photosynthesis seems to violate the Second Law

Carbian About 400 years ago	
Agrian About 12,000 years ago	
Pyrian About 0.5 million years ago	
Oxian About 2.1 billion years ago	
Photian About 3.8 billion years ago	
Thermian About 4.2 billion years ago	

because it spontaneously transforms chaos (widely dispersed water and carbon dioxide molecules) into order (living structures in the form of blue-greens or green plants), thus representing entropy consumption instead of entropy production. Yet, the entropy consumption is local, at the photosynthesis site. Overall, in the global environment, entropy is produced. In energetics jargon a photosynthesiser is a so-called *energy-dissipating structure* which is maintained by an energy flow, in this case a solar energy flow. In effect, the energy flow keeps the energy-dissipating structure far from thermodynamic equilibrium, a prerequisite for life – as the equilibrium state is lifeless.

MODELLING THE FUTURE

One way to explore the unknown is through laboratory experimentation. But when the unknown is 'the future', this method cannot be applied and an alternative approach has to be adopted, that is exploring the unknown through modelling. Devising a model looks like a simple proposition, but the concept has various meanings in different academic languages. Two main types of models can be discriminated: the mathematical and the conceptual.

In many natural sciences, an intellectual construct only earns the qualification 'model' if it has quantitative, predictive power. Hence a system can be modelled only if represented mathematically. To incorporate uncertainties as well as to keep computations manageable – scientists demonstrate a great appetite for computational modelling – simplifications and assumptions are made in describing the real world. Yet, because of the lack of knowledge about their defining rules, mathematical representations of genetic and cultural developmental processes *"will have very limited value"* [69d]. Therefore, the evolutionary model proposed in this chapter, The Staircase of Energy Regimes, is a conceptual model. In the generic sense a conceptual model is a simplified, schematic representation of a system's high-level organising principles and takes into account key components, relationships, and processes.

These days computer models seem to have the primacy. Indeed, they help to deal with immensely complex sets of micro-, meso- and macroscopical data, and to quantitatively describe all sorts of progression. However, extensive computer models covering 'system Earth' are so complicated that only experts understand their output. Moreover, in view of the inherent unpredictability of evolutionary trajectories, 'system Earth' computer models are particularly suitable for short- and possibly medium-term planning. Conceptual models, on the other hand, not only facilitate the evaluation of long-term connections and pathways to the future, but also enable societal and strategic dialogue as well as education.

The Staircase of Energy Regimes

All living structures exist and grow far from thermodynamic equilibrium; thus every living organism or living organisation, from a microorganism to a human being, or from an ecosystem to a sociosystem, dissipates energy. Life emerges and is sustained on the wings of an *energy flow* springing from an *energy gradient* between an *energy source* and an *energy sink*; equally, life will become extinct when the life-giving energy flows fade. Energy flows tend to *drive* the evolution of living structure along a trajectory of increasingly complex so-called *dynamic steady states*, or *attractor states*. The *shape* of an energy-dissipating structure depends on the nature of the interactions, or the *shaping forces*, between its composing parts. A driving force thus results from an energy flow *through* the system, while a shaping force results from the interactions between agents *within* the system.

Intriguingly, an ecosystem or a sociosystem appears to be able to 'sense' energy gradients other than the main one which produces the energy flow that keeps it alive. In addition, when the right evolutionary conditions occur, the system appears to possess the capacity to develop new source-to-sink dissipation paths utilising a newly 'sensed' energy gradient. While actual conversion proceeds via physical, chemical or biological processes, ecological and societal organisations create food chains or energy chains, respectively. Just like biological species, energy systems fight an evolutionary battle. And the leading ones rule the ecologically dominant energy regime, or energy economy. The evolutionary lineage of energy-dissipating structures discovered during the journey reported in Part I has revealed six successive ecologically dominant energy regimes, as illustrated in Figure 6.1. The key observations are described below in terms of evolutionary energetics (discussed in the Appendix) for each energy regime.

The Thermophilic Regime

During the Thermoic, the first energy era on planet Earth, it was heat that drove the processes of change. Heat propelled the chemical evolution or the increase of molecular diversity. Stuart Kauffman suggests that chemical evolution induced the appearance of autocatalytic processes, whereby products of chemical reactions start to act as catalysts for their own production. It is possible that protocells emerged, sustained by an autocatalytic metabolism which, in turn, formed proto-ecosystems: the protocells absorbed molecules from their environment, excreted molecules into their environment and exchanged molecules between each other [14b]. The prevailing view is that life originated in oceans, heated by volcanoes, hot springs and asteroidal impacts, with the emergence of hyperthermophilic microorganisms. These heat-lovers thus built the first step of the Staircase of Energy Regimes, the Thermophilic Regime.

Energy – Engine of Evolution

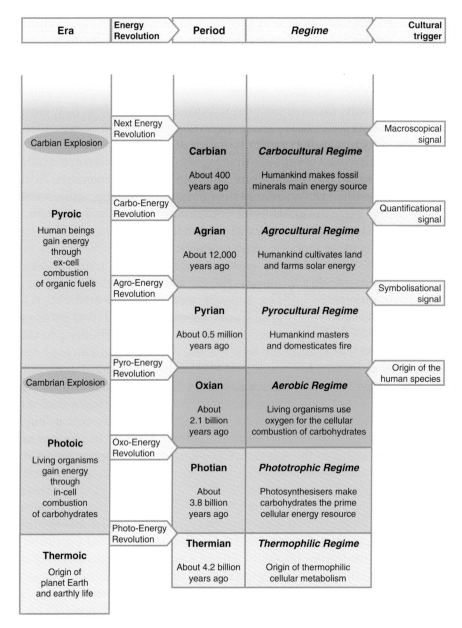

Figure 6.1   Energy time scale of the history of the Earth

Because the first living energy-dissipating structures were hyperthermophilic, the primary *driving force* behind their origination must have been the heat flow created by the energy gradient between the hot earth and its cold atmosphere. That heat flow pushed the chemical systems – often referred to with the colloquial term 'chemical soup' – far from thermodynamic equilibrium. Under such conditions, as modern energy science demonstrates, energy-dissipating structures in the form of autocatalytic cycles can spontaneously emerge to accelerate entropy production, just like a whirlwind in 'hot air'.

An energy-dissipating structure represents information, information that is embodied in the structure itself – in fact, structure *is* information. In a protocell, the information is encoded in the molecular structure of the autocatalytic species. Later, with the genetic innovation, information about biological cell structure – including its metabolic design – appeared in the form of molecular codes written in distinct molecules, or DNA strands. These chemical blueprints most likely accelerated cell division, so increasing entropy production and boosting 'global' energy dissipation.

The Phototrophic Regime
Hyperthermophyles originated around oceanic vents of superheated water called black smokers [31a], but a number of colonies must have reached the water surface at a certain moment in time; maybe millions, or tens of millions, or hundreds of millions years after the origination of their ancestors. Near the water surface, these hyperthermophyles became exposed to incident sunlight, whether they liked it or not. Some of them succeeded in exploiting the incoming energy flow, so giving birth to a new metabolic phenomenon, photosynthetic life. The innovative microorganisms became solar-energy-dissipating structures, and the resulting ecosystems solar-energy-dissipating superstructures. Photosynthesisers established the first solar energy economy on Earth. They dethroned the hyperthermophyles from their ecologically dominant position and launched the Phototrophic Regime.

The *driving force* of the new energy economy was a flow of incoming solar energy. The *shaping forces* originated from physicochemical interactions such as those between solar energy and light-sensitive pigments, as well as biological processes such as natural selection. Of course, a well-documented journal of the evolution of photosynthesis is lacking. However, two key stages can plausibly be reconstructed from paleo-biological analyses. The first stage was the technology to fix carbon dioxide and the second one the technology for stripping the hydrogen required from water. Cyanobacteria, or blue-greens, excelled in this dual solution and, as a result, emitted molecular oxygen, the leftover from hydrogen-stripped water. The utilisation of solar energy freed life from its sole dependence of energy provision by hydrothermal sources. The new energy source enabled an increase in organic productivity by perhaps more than two to three orders of magnitude [41]. By producing carbohydrates from carbon dioxide and water, the oxygenic blue-greens 'charged' as it were this energy sink with solar energy. In effect, they built up a new energy gradient on Earth, which had its origins in the Sun's hydrogen-helium source-sink system.

Biologists talk enthusiastically about evolution's incredible creation of *diversity*, whereas energeticists like to point to the amazing chemical *uniformity*. The hyperthermophyles had selected nucleotides as building blocks for DNA polymers; amino acids as building blocks for proteins; and adenosine triphosphate (ATP) as fuel for cellular work [17c]. The photosynthesisers adopted these hyperthermophilic 'standards', which did not change in more than 3.8 billion years of biological evolution. Still, manufacturing ATP from photosynthetic, energy-rich sugars is not too easy a task. Fermentation, an early sugar combustion technology, yields approximately two ATP molecules per glucose unit (a well-defined sugar) [34d]. This amounts to about 5 percent energy recovery based on the 'photosynthetic crude' [35d], an efficiency leaving room for improvement.

The Aerobic Regime
The blue-green's evolutionary success unchained a positive feedback loop between population size and oxygen build-up in the environment. More solar energy utilisation led to more photosynthesising organisms, yielding more hydrogen-stripping from water, and thus more oxygen being released. After innumerable emitted oxygen molecules had 'oxidised the Earth', the blue-greens started to enrich the atmosphere with this colourless, odourless and tasteless gas. Life created a photochemical oxygen pump driven by the Sun.

But oxygen was a poison to almost every living thing, including the ruling blue-greens themselves. Their way of living became unsustainable. But with the dramatic environmental change a new energy gradient arose embodied by the source-sink system of photosynthetic carbohydrates plus oxygen, and carbon dioxide plus water. However, without conversion paths a new energy gradient is useless. Reversed photosynthesis would be an obvious option, but far-from-equilibrium processes are irreversible; the Second Law forbids them, allowing forward policies only. New pathways had to emerge for the conversion of photosynthetic carbohydrates plus oxygen into carbon dioxide plus water. Certain bacteria 'sensed' the new energy gradient and invented *aerobic respiration*, a truly ground-breaking bio-combustion technology: aerobic respiration produces 18 times more cell fuel, or ATP, from crude sugars than anaerobic phototrophs.

The Staircase of Energy Regimes

But the oxygen-breathers could not harvest sunlight by themselves. They were fully dependent on the blue-greens' sugar production. Thus with blue-greens struggling for life in their own oxygen waste, the bacterial aerobes carrying state-of-the-art combustion technology ran the risk of becoming extinct, too. A strategic alliance, or a symbiotic partnership, between the oxygen producers and the oxygen users overcame the deadlock. The resulting new cellular design had it all: sunlight-harvesting photosynthesis, highly efficient metabolic combustion through aerobic respiration, and also protection against oxygen toxicity. In terms of evolutionary energetics, the symbiotic alliance gave birth to new energy-dissipating structures, or eukaryotic cells. The revolutionary eukaryotic energy utilisation enabled the emergence of multicellular aerobes driving life forms and ecosystems to higher complexity. The eukaryotic aerobes triumphed over the blue-greens and established the next ecologically dominant energy regime, the Aerobic Regime.

The Pyrocultural Regime

The Aerobic Regime reshaped planet Earth. Nature underwent unprecedented change. Plants and animals occupied barren land, the once oxygen-free atmosphere became enriched with about 20 percent of this 'strange' molecular species, and last, but not least, wild fire entered the biosphere. Just as happened twice before in the history of life, a new energy gradient emerged: the source-sink system of wood plus oxygen, and carbon dioxide plus water. From nature emerged two conversion paths: rotting (by decomposing microorganisms) and burning (by wild fire). Then genus *Homo* recognised the energy gradient and invented a revolutionary new one: collect wood, catch fire by torch, light the wood-fuel and maintain the fire. He tamed wild fire so augmenting his 'energy menu' substantially through applications such as heating, lightning, roasting and scaring away animals. Says Johan Goudsblom, fire mastery marks the origin of human civilisation.

The emergence of anthropogenic fire heralded the origin of a new energy-dissipating structure, the pyroculture. Pyrocultural life descended from aerobic life, which had built the energy gradient driving it. The new energy-dissipating structure was not biological, nor ecological, but societal by nature. In effect, fire mastery went with the creation of a 'socio-chemical' metabolism, with the fireplace as the first cultural – not biological – metabolic site. Just like physical, chemical, biological and ecological far-from-equilibrium structure, societal structure needs an *energy flow* of high-quality energy to emerge and sustain, or to live – in this case heat and light from burning wood. The process is accompanied by entropy production in line with the Second Law as reformulated by Eric Chaisson (see the Appendix): local order, such as societal order, must be compensated with global disorder, or 'chaos', through energy dissipation. And just like the earlier energy-dissipating structures, the fire economy tended to move further away from equilibrium during the course of time, continuously developing dissipative paths and catching more and more energy.

A characteristic of energy-dissipating structures is their capacity to spontaneously develop new structures or *emergent properties*. A whirlwind is a physical emergent property of hot air, a species is a biological emergent property of the biosphere, and a population distribution is an ecological emergent property of an ecosystem. As the fire economy was a societal dissipative structure, its corresponding emergent properties were cultural by nature. For example, 'sitting around the fire activities' were *emergent properties* of the new pyrocultural energy-dissipating structure. As fire mastery forced people to coordinate and communicate as never before, speaking language, thinking symbolically, and developing human consciousness and intelligence are strong candidates for being pyrocultural emergent properties. Hence the selection pressure to evolve fire mastery must have been high, as it improved both the chances of survival and the quality of domestic life.

What is true for energy-dissipating structures in general holds for fire mastery, too: stop the energy flow and the corresponding structure will collapse. The *driving force* of pyroculture was a flow of light-giving heat from burning firewood; thus converted solar energy. But for the first time in the history of life, the prime *shaping force* of an energy-dissipating structure sprang from human ingenuity. Fire masters created technological artefacts such as the torch and the fireplace to conduct their new energy economy, the fire economy, which soon established ecological dominancy. Pyroculture happened to be a very successful evolutionary strategy.

The Agricultural Regime
At a certain moment in time humans began to deliberately burn fields, sow seeds from edible crops, and harvest self-produced food. Natural growth became cultural, or rather agricultural. In terms of food provision agriculture is orders of magnitude more efficient than hunting and gathering. It emerged from the foundation laid down by pyroculture, which yielded the fire technology to control the burning of fields, and to cook the harvested crops. Of course, a burned field is not an energy source, but in combination with seeds, carbon dioxide and water it is an energy sink: fallow land can, as it were, be 'charged' with solar energy to yield an energy source. Just like the photosynthesisers billions of years ago, the solar

farmers succeeded in deriving a new type of energy flow from the Sun, in this case an agricultural flow of food. The selection pressure on farming grew with rising food scarcity, which partly resulted from its own success.

Agroculture, like pyroculture, is a societal energy-dissipating structure. It evolved along a trajectory of complexifying societal states, utilising constantly more solar energy. The agricultural food flow became the *driving force* of human development, *shaped* by creative people and their self-made tools. Numerous new *properties emerged* from the agricultural energy-dissipating structures, among which are crafts, villages, a growing population and new energy chains. Cooking 'escalated' from preparing the meal to making glass and brick, to ironworks. And just like with the evolution of biological species, agricultural evolution produced uniformity *besides* diversity, as is evident in emergent properties such as the alphabet, time and money.

The heat released from burning wood found so many applications that forests disappeared at a faster rate than nature could renew. Agrocultural Man was a fire master. Although other energy systems, such as wind power and waterpower, emerged, and did play an important role in the move up to the next energy regime, none of them attained ecological dominance.

The Carbocultural Regime
Long before the origin of genus *Homo*, in the Aerobic Regime, huge volumes of dead biomass ended up in deep layers of the Earth's crust. And, depending on both the nature of the biomass and the geochemical conditions of the 'burial place', either coal, petroleum or natural gas was created by a process of fossilisation. The biogenic carbohydrates changed into mineral hydrocarbons, which led to the origin of a substantial energy gradient between fossil fuel plus oxygen, and carbon dioxide plus water. Only conversion paths needed to be developed.

It was *Homo sapiens carbonius* who finally recognised the fossil energy gradient and managed to utilise it for his own benefit. Through the exploration for and the production of fossil fuels, he unlocked the massive reserves of 'ancient sunlight'. First Carbocultural Man replaced firewood and charcoal with mineral coal, particularly in 'industrial cooking'. But the real break-through was burning a fossil fuel in a so-called combustion engine to convert heat into mechanical power for a wide variety of purposes; think of a pump to distribute water, a vehicle to move goods and people, or a dynamo to generate electricity.

A flow of heat from burning ancient biomass has become the *driving force* of a new energy-dissipating structure, the carboculture. Carbocultural Man is a fire master as well as a solar farmer, but lives largely off the energy-rich

Energy – Engine of Evolution

remnants of ancient aerobes. The stream of fossil fuels unearthed has yielded an unprecedented *energy flow* of heat through unprecedented societal energy-dissipating structures, with their associated unprecedented *emergent properties* (electricity, quantum mechanics, antibiotics, pop music, the world-wide web, man on the moon) and unprecedented growth rates of population and economies. Also, within these carbocultural energy-dissipating systems, increasingly complex societal structures have emerged (cities, the United Nations, unions, the World Wide Fund For Nature) as well as enormous artificial order (buildings, vehicles, medicines, computer networks, mobile phones). This explosion in order has been paid for by a concomitant explosion in the utilisation of low-entropy fuels and a corresponding explosion of produced entropy in the form of rejected heat plus material emissions. This Carbian Explosion is one of the most impressive events in the history of life on Earth. Carboculture has grown to be ecologically dominant in record time.

Homo sapiens carbonius, energy-hungry as he is, continuously hunts for new energy gradients to exploit. Several 'carbon-less' energy chains have already appeared, for example nuclear fission and photovoltaics. Moreover, scientists have 'sensed' the energy gradient embodied by the source-sink system of heavy water plus lithium and helium (plus energy-rich neutrons); they now search for feasible conversion paths to mimic the Sun on Earth in a man-made nuclear fusion reactor.

Building the Staircase of Energy Regimes is rather straightforward once the foundations have been laid. The Staircase simply uses stairs to represent the ecologically dominant energy regimes that occurred during the history of planet Earth. Because each energy regime added a new form of complexity to the living planet, the result is a staircase rising with time, as illustrated in Figure 6.2.

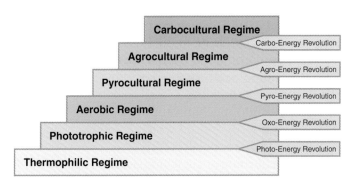

Figure 6.2 Staircase of Energy Regimes

THE EVOLUTION OF ENERGY AND ENTROPY FLOWS

Ever since the origin of photosynthesis, more than three and a half billion years ago, biological and cultural life has been driven by energy from the Sun, albeit packaged in different forms: fire, food and fossil fuels. Figure 6.3 illustrates the lineage of energy regimes during the evolution of life on Earth, with their energy gradients, typical conversion paths, and corresponding energy flows.

Five times a new energy regime was developed from its 'parent regime' by devising a novel conversion path for a new energy gradient (which could have emerged in an earlier regime). Notably, energy regimes not only evolve themselves as eco- or sociosystems, but they also prepare the build-up of new energy gradients through the production of energy sources and/or sinks.

The Thermophilic Regime produced heat-loving bacteria, some of which became exposed to incident sunlight. That opened the way for the development of chemical paths to exploit the flow of incoming solar energy, which resulted in photosynthesis. Photosynthesising bacteria evolved a route to reduce carbon dioxide with hydrogen stripped from water, which yielded free molecular oxygen – a toxic for almost all that lived under the Phototrophic Regime. But free oxygen plus photosynthetic carbohydrates introduced a new energy gradient with carbon dioxide and water as an energy sink. Aerobes succeeded in utilising the fresh energy gradient through aerobic respiration, a cellular *in vivo* combustion technology. The Aerobic Regime realised a fabulous growth in biomass, which largely ended up as fossilised minerals. Human beings, relatively intelligent and recent aerobes, invented a pathway to utilise the energy gradient between inedible biomass plus oxygen and carbon dioxide plus water. Source-to-sink conversion went through fire mastery. That yielded artificial fallow, which could be combined with seeds to harvest solar energy through growing edible crops. From agricultural *energy-dissipating sociosystems* emerged the knowledge and the technology to exploit the enormous and long-standing energy gradient between fossilised biomass plus oxygen and carbon dioxide plus water. *Homo sapiens carbonius* developed the conversion technology still further with the combustion engine.

Except for the first energy regime, the Thermophilic Regime, all five subsequent energy regimes derive their energy flow from a flow of solar energy. Hence the five earthly energy gradients which have been driving life on Earth since the origin of photosynthesising organisms are actually descendants from the energy gradient between hydrogen and helium exploited in the Sun by nuclear fusion. Note the central play in the energetics of life of 'charging' carbon dioxide plus water with sunborne energy, and then 'discharging' the resulting carbohydrates.

> **THE EVOLUTION OF ENERGY AND ENTROPY FLOWS (CONTINUED)**
>
> The scheme of evolutionary energetics presented here forms the justification and foundation for the Staircase of Energy Regimes. Other 'energetic species' have been originated, particularly from the Agricultural Regime and the Carbocultural Regime (examples are wind, solar, hydro- and nuclear power), but they did not conquer ecological dominance in an energy period.

The rise and fall of complexity

In his fascinating chronicle 'Steps Towards Life: A Perspective on Evolution', biochemist and Nobel laureate Manfred Eigen notes that complexity has accumulated throughout biological evolution from the first single-celled organisms to human beings [32b]. Astrophysicist Eric Chaisson expands on Eigen's observation in his book 'Cosmic Evolution: The Rise of Complexity in Nature'. He argues that the information content of the Universe as a whole has been increasing since the Big Bang, from particles to atoms, to galaxies, to stars, to planets, to life, to intelligence, and to culture [36b]. Historian Joseph Tainter, in focusing on human civilisation, points to the rise of complexity of human societies in going from hunting and gathering to modern industrialised structures. Notably, both Eigen and Tainter implicitly define the concept of complexity in line with Chaisson: as *"a measure of the information needed to describe a system's structure and function"* [36b]. So considered, the words *information* and *complexity* also refer to notions like *order*, *structure*, and *organisation*.

In his quest for the fundamental drivers of the complexity increases during cosmic evolution [36b,c], Chaisson concludes that [36d]:

> *"More than any other term, energy has a central role to play in each of the physical, biological, and cultural evolutionary parts of the inclusive scenario of cosmic evolution; in short, energy is a common, underlying factor like no other in our search for unity among all material things."*

To be precise, it is not the energy content of a structure, but the energy flow through it, that drives the structural changes [36e]. Chaisson argues that during cosmic evolution not only the complexity of structures, but also the energy flow through these structures has continually increased in time [36f]. He claims that the close relationship between the rise of *structure complexity* and the rise of *energy flow through structures* is so universal and so fundamental that *"the two are the same, or nearly so"*, indicating that

The Staircase of Energy Regimes

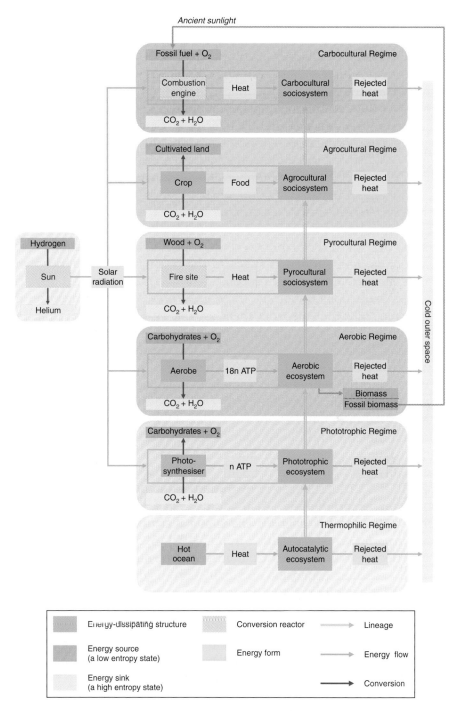

Figure 6.3   Evolutionary energetics

'complexity' and 'energy flow' have been continually rising with time since the origin of the Universe. Yet zooming in on energy regimes reveals considerable collapses of complexity.

For more than three billion years the complexity of life developed progressively from the first hyperthermophyle communities to the eukaryotic richness after the Cambrian Explosion (about 535 to 525 million years ago) – not continuously, but persistently over time. Since then the number of biological genera – a measure of biodiversity, and thus the complexity of life – roughly quadrupled, although some biologists challenge that view: they conclude, tentatively, that the number of genera over the past 450 million years, on average, remained about constant, suggesting *"there may be an upper threshold to biodiversity – a maximum holding capacity of the environment"* [101]. However, the paleo-record reveals five pulses of catastrophic mass extinctions over the last 500 million years, triggered by, for example, volcanoes, asteroids, and sudden changes to the oceans and the atmosphere. Each time life recovered, albeit probably rerouted onto a new set of tracks [28g]. Mass extinctions divide the three great eras of the last aeon on the geologic time scale: the Paleozoic, the Mesozoic and the Cenozoic ('ancient life', 'middle life' and 'early life', respectively). It is alleged that 99.9 percent of all species have come and gone [14c]. During the Permian extinction (about 250 million years ago), over 90 percent of all species on Earth disappeared, causing the world's ecosystems to collapse *"as a house of cards"* [28g,w]. Human civilisations show patterns of rise and collapse as well: the first Mesopotamian Empire, the Egyptian Old Kingdom, the Roman Empire and the Lowland Classic Maya emerged and became extinct.

Perhaps living systems, once emerged, continually attempt to increase in complexity,* but this is not the same as continually succeeding in that attempt. In the various stages of evolution, extinction is just as normal a phenomenon as origination. Also today *"rapid population declines and extinctions of species following the widespread destruction of natural habitats have been reported across the natural world"* [102]. Evolutionary complexity increases, decreases and ceases. Change is the true nature of evolution, and all change – creation, expansion and extinction – is driven by surging, sustaining and fading energy flows. In accordance with the Second Law, evolutionary processes progress irreversibly with time, because they proceed far from equilibrium, and far-from-equilibrium processes are irreversible. Richard Dawkins labels 'progress' as either value-laden or value-neutral, that is *"either with or without building notions of what is good or what is bad"*

---

*Several theoretical ecologists have drawn this conclusion for ecosystems. See, for instance, 'Thermodynamics and Ecological Modelling', in which Sven Jørgensen states: *"Ecosystems attempt to develop toward a higher level of exergy."* [104a]

The Staircase of Energy Regimes

[103]. As the Staircase of Energy Regimes depicts 'thermodynamic progress', it is value-neutral in Dawkins' terms. The Staircase of Energy Regimes can be seen as a one-way Staircase of Evolutionary Progress with three degrees of freedom: the next energy regime might add complexity to the living world (that is upstairs), and it might remove complexity (that is downstairs), but it has no choice than to evolve in a forwards direction – the Arrow of Time is inexorable, but not the Rise of Complexity.

The Staircase is a multifaceted construction: it represents a lineage of complex adaptive systems evolving along a trajectory of attractor states; it represents a lineage of evolving dynamic steady states of energy-dissipating structures maintained by flows of energy; and it represents a lineage of complexifying metabolisms with distinctive energy and material flows, including corresponding structural information.

THE EVOLUTION OF INFORMATION

Generating information

Billions of years ago early photosynthesisers began to capture an infinitesimally small portion of a 'minute' solar energy flow; about one-billionth of the Sun's energy output hits the Earth [39d]. Nowadays still less than one percent of this one-billionth of the Sun's output is captured by 'sunlight-eaters'. Still, this seemingly negligible fraction of energy drives the evolution of life on Earth. The Staircase of Energy Regimes is an account of this evolution from an 'energetical' perspective rather than from a biological or cultural angle.

For today's biologists evolution means 'genetic evolution'. In their view naturally selected genes structure the living world, as Stuart Kauffman contends [14d]:

> "Make no mistake: the central image in biological sciences, since Darwin, is that of natural selection sifting for useful variations among mutations that are random with respect to their prospective effects on the organism. This image fully dominates our current view of life. Chief among the consequences is our conviction that selection is the sole source of order in biology. Without selection, we reason, there could be no order, only chaos."

But Kauffman does not agree with this view [14b]:

> "We are the children of twin sources of order, not a singular source. We have seen that ... the origin of life itself comes because of what I call "order for free" – self-organization that arises naturally."

Kauffman identifies 'self-organisation' as the partner of natural selection in moulding life [14e]. But as 'self-organisation' is energy-driven, 'energy-driven organisation' would be a more precise term. Evolution thrives on the wings of energy-dissipating structures *driven* by flows of energy. A *driving force* that sustains an energy-dissipating structure originates from an energy gradient, whereas *shaping forces* result from interactions between the structure's composing parts, and between these parts and their environment. Think, for example, of the vortex in a bathtub, a physical energy-dissipating structure. The *driving force* is gravitation, while the *shaping forces* emerge from physical interactions between the individual water molecules, and from the interactions between the water molecules and the wall of the bathtub; the shape of the bathtub influences the shape of the vortex. No single water molecule is encoded with information about such a thing as vortex structure. They all interact synchronously in 'their selfish way'. *Strings of synchronised interactions* between the water molecules do the shaping. The vortex structure thus arises from the whole, not from the parts, seemingly shaped by an 'invisible hand'.

Then think of the energy-dissipating ecosphere, which, like an invisible hand, maintains the atmospheric oxygen concentration at around 20 percent. It is a property of the whole energy-dissipating ecosphere, shaped by *strings of synchronised interactions* between oxygenic organisms in their environment of inanimate and biotic factors (such as climate, soil and other living things); no single gene is encoded for atmospheric properties. And think of human societies in which, according to Adam Smith, self-seeking man is often *"led by an invisible hand … [to] advance the interest of the society"* [105]. Adam Smith's invisible hand is a property of an energy-dissipating sociosystem shaped by *strings of synchronised interactions* between human beings in their economic environment; no single human brain is encoded for a trade cycle. Generally, 'invisible hands' can be seen as manifestations of energy-driven organisation.

The nature of a shaping force is *encoded in the structure* of a corresponding shaping agent. For example, the physicochemical properties of a water molecule are encoded in its molecular structure, which was H_2O when planet Earth was born, and will be H_2O when planet Earth dies. Since the molecular structure of water remains constant over time, as does the information encoded, a water vortex always looks the same under the same conditions, being as constant as water's boiling point, yet another ensemble property of very many water molecules. However, the interacting agents of eco- and sociosystems *do* change over time, as information encoded in genomes, brains and artefacts is sensitive to mutagenesis. Hence, bio-, socio- and techno-, or better 'arto-information',* evolves, and thus biological, societal and artificial shaping forces evolve. This is a fundamental difference between a physical energy vortex and 'energy vortices' in living eco- and sociosystems.

For Manfred Eigen, the key word representing the phenomenon of 'complexity' is 'information' [32c]. In his view [32b]:

> *"Evolution as a whole is the steady generation of information – information that is written down in the genes of living organisms."*

Although Eigen talks in the context of biological evolution, his notion generally holds true: evolution as a whole is the steady generation of information – information that is encoded in the genomes, brains and artefacts of living organisms. Information is the common denominator, and represents order, or structure – energy-dissipating structure, to be precise.

The historian David Landes remarks in his 'The Wealth and Poverty of Nations: Why Some Are So Rich and Some Are So Poor' [75a] that:

> *"All economic [industrial] revolutions have at their core an enhancement of the supply of energy, because this feeds and changes all aspects of human activity."*

To generalise again: all biological and cultural revolutions have at their core an enhancement of the supply of energy, because this feeds and changes all aspects of ecological and human activity.

When I combine the above two generalisations it follows that the evolution of life can be seen as one evolution of information, or order, or energy-dissipating structure, driven by flows of energy – which could be enhanced through an energy revolution, whether biological or cultural by nature.

From genes to memes and artefacts

If evolution as a whole is the steady generation of information, what then are the *units of information*? It is widely accepted that the gene is the information unit of heredity in biological evolution. That is not the same as saying that biologists agree about the traits that genes pass down the generations. Some think the genes are responsible for *all* traits, whether the living organism is bacterium, plant, animal or human being, while most make an exception for *"the most cognitively sophisticated animals"* [10d], such as primates, showing cultural behaviour. Recently, though, a number of

---

*The term 'technosphere' is rather common, but I would prefer 'artosphere' with 'arto' refering to human-made. The 'artosphere' is the whole of human artefacts; not all artefacts are technological by nature.

biologists have acknowledged that there is *"a rapidly accumulating body of evidence"* [106] that cultural transmission also occurs *"in what is used to be referred to as "lower" vertebrates"* [10d]. Says Lee Alan Dugatkin [10b]:

> *"We tend to think we are the only animals able to do the trick of passing down the wisdom of our forebears. That trick is known as culture. Surprisingly, even guppies can do it."*

Noticeably, cultural transmission takes place not only between the generations, but also within them [10e]. Hence cultural evolution – understood *"as the summed effect of cultural transmission over long time"* [10f] – proceeds much faster than genetic evolution [10g]. Cultural transmission of information is defined as involving some mix of trial-and-error learning, social learning as a result of observation and imitation, and, in special cases, teaching. Genetic and cultural evolution are two separate entities, interacting in co-evolutionary and sometimes bizarre ways [10a]:

> *"These two forces may act in the same direction, or they may conflict with one another. Interactions in which genes or culture or some combination of the two win the day are all possible. Genes may even code for culture, but the manifestation of such culture – for example, fads and craze – cannot be measured in any meaningful way by studying genetic architecture."*

But if it is not genetic architecture, which architecture is it? What are the units of information by which culture is encoded? In 'The Selfish Gene' (1976) Richard Dawkins coined the name *meme* for the unit of cultural inheritance [107a]. Later, in 'The Extended Phenotype: The Long Range of the Gene' (1982), Dawkins writes he was initially *"insufficiently clear about the distinction between the meme itself, as replicator, on the one hand, and its "phenotypic effects" or "meme products" on the other"*, continuing that [108a]:

> *"A meme should be regarded as a unit of information residing in a brain. It has a definite structure, realized in whatever physical medium the brain uses for storing information. If the brain stores information as a pattern of synaptic connections, a meme should in principle be visible under the microscope as a definite pattern of synaptic structure. If the brain stores information in 'distributed' form, the meme would not be localizable on a microscope slide, but still I would want to regard it as physically residing in the brain. This is to distinguish it from its phenotypic effects, which are its consequences in the outside world."*

The idea of the meme induced a lot of turmoil among socio-biologists and social scientists. Clearly, Dawkins' original definition still lives on, fighting a memetic struggle with its refined description. Dugatkin adopts the reformulation [10h], but concludes that *"after all, the meme still plays second fiddle to the gene in Dawkin's orchestra of life"* [10a]. Indeed, Dawkins introduced the meme in the first place to explain human, not animal, culture [107a].

In his book 'The Electric Meme: A New Theory of How We Think', anthropologist Robert Aunger further develops memetic theory. Aunger defines the meme in terms of a physicochemical *state* somewhere in the brain [109a]: whereas a gene is encoded by the molecular structure of a DNA molecule, a meme is encoded by the electrochemical state of supermolecules in the brain cells. In effect, the information sits in electrically polarised nodes in the brain's neural network [109a,b]. Compare this with the hard disk of a computer: no material is added to the hard disk as data are accumulated, just as no material is added to the brain during learning [109c]; ideas are weightless, even if great. Ideas can be seen as *expressions* of patterns of brain states, like a picture on a computer screen is an *expression* of a pattern of 0's and 1's generated by the computer, and the colour of an eye is an *expression* of a molecular pattern programmed in genes (and as the screen picture is no bit, the eye no gene, that beautiful melody is no meme).

According to Aunger, memes can have a genetic cause, or a 'natural' bias, which is consistent with the notion of significant innate structuring of the brain. In addition, they can emerge in the brain through spontaneous mutation as a result of internal mental activity; after all, the brain is an energy-dissipating structure in itself with inherent self-organising, or rather energy-driven organising, capabilities. And thirdly, the origination of memes can be induced by signals from the brain's environment [109d]. Social transmission results in the mingling of ideas, beliefs and values, and thus human culture [109e]. Hence memes are inherently social [109f]. They fight an evolutionary battle, competing for the right to occupy a 'mental niche' or 'cultural niche', with 'mental selection' [109g] or 'cultural selection' doing the sifting. Humans store and maintain cultural information in memes, but not only in memes. Memes can be expressed in the form of *artefacts*, and artefacts could in turn be designed to spread expressions of memes. Think of mural painting, papyrus leaves and the printing press that multiplies a communicative artefact such as a book [109h]. Aunger views technological innovation as artificial evolution, or *"a Darwinian process of descent with modification operating in parallel between ideas and their implementation"* [109i]. *"Memes must dance a tango with technology"* [109j], he says.

Culture has a body and a mind [109k], the 'artosphere' and the 'noösphere';* all that humans create, from pencil to painting, from wheel to

---

*From the Greek word 'noös', which means 'mind' [110].

wagon and from chip to computer, constitute the artosphere; and all shared human ideas, values and beliefs, all that humans think together, builds the noösphere or the 'mindsphere'. The artosphere, or the sphere of human artefacts, and the noösphere, or the sphere of human memes, together form the anthroposphere, or humansphere. Our artificial and memetic achievements tend to catch the eye in evolutionary history, but let us not forget the biological groundwork that provides our genus with handy bodily manipulators such as a good voice box for advanced social communication.

Human culture has been a co-evolutionary achievement of genes, memes and artefacts since the origin of human culture about 2.5 million years ago, when Man the Toolmaker arrived. After millions of years the human body reached its modern 'steady-state structure', and human culture began to evolve at an unprecedented rate. Memes and artefacts started a co-evolutionary rat race, leaving genes behind in the dust. The Pyro-Energy Revolution, as much a societal as an artificial event, enforced the co-evolution of memes and artefacts, and thus the evolution of human culture. During the Agrian Period humans began to artificially select natural crops, and currently carbocultural memes convert genes into artefacts (in a metaphorical sense, of course). The symbiotic cooperation between human memes and artefacts has grown into an ecologically dominant shaping force, driven by surging flows of fossil energy manipulated with advanced technological artefacts. Whereas today, the pools of human memes and artefacts are exploding, the gene pool is shrinking as shown by the loss of biodiversity. Memetic and artificial evolution added new lines of development to the genetic phenomena – particularly for human beings, whose power to represent symbolic relationships *"created a whole new landscape for the mind to wander, free from the bounds of the immediate present and possible"* [56m]. Memetic processes allow human beings to imagine, to fancy, to think the unthinkable.

SOCIO-TECHNOLOGICAL DEVELOPMENT

The rhythm of societal development

At this stage the Staircase of Energy Regimes is a rather coarse construction. To better understand societal development we must invoke resolution. This requires us to zoom in on the anthropic era, the period that has seen the evolution of genus *Homo*, and thus the anthroposphere. In the anthroposphere the following 'anthroposystems' can be distinguished: *human knowing*, *human capacity*, *human acting* and *human living*, with the

The Staircase of Energy Regimes

corresponding developmental vectors *discovery, invention, innovation* and *diffusion*:

Discovery : is, as the word says, dis…covery: an act of removing the 'cover' from *something that exists*. It is a first-time *observation* of an object, a phenomenon, a pattern, or a quality of natural or cultural origin, thus the observation of a new fact or event. A discovery stimulates and enables the development of thoughts, hypotheses, explanations, insights and theories – in short, knowledge.

Invention : is the creation of a new capacity based on new knowledge. An invention is something that did not exist before and enables the development of applications in the form of new tools or techniques. It could be an artefact or it could be a method or a model.

Innovation : is the successful, practical application of a newly invented tool or method. In business, this means bringing insightful ideas successfully to the market [111a].

Diffusion : is the expansion of a new market, or the spread of a new product through a community. It is the movement from niche to commonplace, which enables the development of new *ways of living*.

These four vectors, as well as their mutual relationships, can easily cause confusion; in particular the word 'innovation' has many meanings. For some it means 'renewal' in general and then all development is innovation; one comes across terms like 'methodological innovation' and 'social innovation'. And whereas academic researchers talk about 'technological innovation' [112a], business people carefully discriminate between 'invention' and 'technology-enabled innovation', aware that patenting costs money and innovations have to make it in the marketplace.

Human culture evolves a universal four-beat rhythm, the rhythm of *discovery – invention – innovation – diffusion*. There cannot be diffusion without preceding innovation; and there cannot be innovation without a preceding invention; likewise, there cannot be invention without a preceding discovery. This holds true for both technological and non-technological progress. Hence, the four-beat rhythm reveals another, more detailed staircase, the Staircase of Socio-Technological Development, with 'socio' referring to shared brainpower, or memetic information, and 'techno' to the man-made world, or artificial information. The steps of the Staircase of Socio-Technological Development add an extra layer to the

Energy – Engine of Evolution

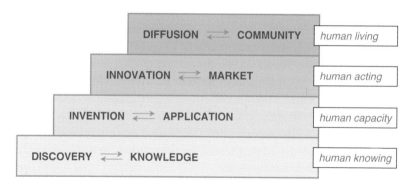

Figure 6.4   Steps of socio-technological development

Staircase of Energy Regimes. In effect, the four socio-technological steps together (see Figure 6.4) represent the human culture or civilisation of a corresponding energy regime. I will now write the rhythms of socio-technological development for each energy regime and build the corresponding Staircase of Socio-Technological Development based on observations made during the time travel in Part I. Note that the four-beat rhythm when slightly transposed sounds like *new observation – new creation – new practice – new way of living*.

The Aerobic Regime

The anthroposphere emerged with the origin of human culture about 2.5 million years ago, at the very end of the Aerobic Regime. Early people *observed* the hardness of stone and the sharpness of some edges. They *created* stone tools such as axes, cleavers and picks by striking stone against stone. Then they used the tools in various *practices* [55d], and their *lifestyle* changed accordingly. These people primarily used artefacts to prepare food, "*substituting stone flakes for the long slicing teeth that cats and other carnivores employ to strip meat from a carcass*" [54c]. Man the Toolmaker, a cave dweller [31f], survived through earth wisdom, or expert knowledge of his natural surroundings [39e]. Thus the first steps of the Staircase of Socio-Technological Development can be characterised with the terms *earth wisdom* for human knowing, *stone technology* for human capacity, *foraging and scavenging* for human acting and *cave dwelling* for human living.

The Pyrocultural Regime

During the Pyrocultural Regime anthropogenic fire mastery emerged. Human beings *observed* the heat of wild fire and the burning of wood.

The Staircase of Energy Regimes

They *invented* the torch as a tool to capture fire, and *created* the fireside to sustain it. They began to use the hearth in daily *practice*, so making cave dwelling more attractive. When the fireside became a common feature, the *lifestyle* of the fire masters changed markedly. They also *observed* half-charred animals and fruits, which tasted better, or became edible. By imitating 'wild' roasting, they *invented* cooking and food conservation. The cooking *practice* extended people's food menu and stimulated the cultivation of eating habits. Conservation through cooking enabled food storage, which allowed populations to stabilise at higher densities [33e]. When cooking became a common habit, a new *lifestyle* emerged. The fire masters also *observed* the fear of fire that other animals had, *created* fire techniques to chase away competitors and *applied* these to improve their chances of survival. The use of fire to fight competitors became *common*.

Fire masters lived in groups of about 100 to 250 people with strong family ties and complex social rules. Women gathered plants and the men hunted [113a]. Except for some maritime cultures, most groups lived a nomadic life so as to find enough food and firewood for survival [33e, 62h]. Within a band people worked together and groups as a whole were self-sufficient. Around the fireside a process of civilisation evolved as people domesticated fire – the fireside became the centre of social evolution. Social evolution, in turn, probably induced the most remarkable development of the uniquely human capacity to think symbolically. This induced the rise of modern language and a new way of thinking about nature and themselves. With fire mastery, the co-evolution of brainpower and technique, or memes and artefacts, accelerated markedly. Near the end of the Pyrocultural Regime, around 50,000 to 40,000 years ago, a sudden growth of symbolic activity, in the form of 'non-useful' artefacts such as jewellery, ornaments and musical instruments as well as graves, coincided with the arrival of *Homo sapiens*. Ideological or religious symbols [54r] indicate the origination of ethnographic 'cultures' or identity-conscious ethnic groups [54s]. The outbreak of symbolic 'doings' can be seen as a Symbolisational Signal (see Chapter 3).

The single key words to characterise the anthroposystems of the Pyrocultural Regime are *symbolic thinking* for human knowing, *fire technology* for capacity, *hunting and cooking* for acting and *nomadic bands* for lifestyle.

The Agricultural Regime

During the Agricultural Regime people became solar farmers. They *observed* how nature reacted to fire and *discovered* the relationship between burned land and plant growth: new vegetation sprouted, which in turn attracted grazers. This triggered the *invention* of agricultural techniques to control plant growth with fire and sowing. *Innovative* plant cultivation led to an increase in food production. With crop cultivation, actually a form of domestication, early farmers settled down; they exchanged a nomadic way of life for a sedentary *lifestyle*.

The use of clay as a construction material for hearths must have led to *observing* its hardening with heating, a *discovery* enabling the *invention* of pottery-making. Several *innovations* appeared, from pot, to dish, to oven [15p]. *Applying* the oven improved cooking. Storage of grain, nuts and other food became a *practical* use of pottery that increased shelf time. Pottery changed the *lifestyle* of people.

Goudsblom writes that the attention farmers gave to cooking yielded the first subtle and intimate knowledge of matter, and thus became the foundation for the empirical sciences [15i]. People must have *observed* material changes like melting and solidification, which happens, for example, when ochre is heated for pot decoration (ochre contains iron that changes colour when heated) [15p]. This, then, could have inspired keen investigators to *invent* metallurgical techniques, which in turn enabled the development of a wide range of *innovations* in the form of agricultural and other implements, such as cooking utensils, ornaments and weapons. To an even stronger extent than for pottery, metallurgy required social co-ordination and specialised know-how; for instance, to guarantee energy supply. It is an understatement to say that the achievements of metallurgy, after their *diffusion* through communities, brought about many radical *lifestyle* changes. Finally, the increased economic diversity in goods and services induced barter trade which became the dominant economic paradigm.

Most remarkably, during the Agrian Period humankind developed the uniquely human quality 'to measure reality'. As happened towards the end of the Pyrocultural Regime, a strong signal surged around circa 1250 to 1350 near the end of the Agrocultural Regime: this is termed the Quantificational Signal. And, just like the Symbolisational Signal, the Quantificational Signal triggered a radical change in people's perception of reality. It touched all anthroposystems in their essence, prompting "*modern science, technology, business practice, and bureaucracy*" [16e]. The late agrocultural acceleration of the co-evolution between brainpower and technique, or memes and artefacts, gave a lift up to the next energy regime, the Carbocultural Regime.

For characterising the four anthroposystems of the Agrocultural Regime, the key words are *practical know-how* for human knowledge, *agricultural technology* for capacity, *farming and bartering* for acting and *farms and villages* for ways of living.

The Staircase of Energy Regimes

The Carbocultural Regime

The Carbocultural Regime gave rise to the Carbian Explosion, during which the anthroposphere, and thus its four anthroposystems, 'exploded'. The agrocultural Quantificational Signal had fired up both the emergence of modern, reductionistic science, and also, but independently, of methodical advancements in the crafts. It took some time, though, before scientists and craftsman chose to ally, but in the end they did. The science-crafts alliance gave birth to modern technology, that is technology based on a fundamental knowledge of natural phenomena. When the alliance succeeded in exploiting 'ancient sunlight', the carbocultural symbiosis began to grow the fossil-fuelled industries as we know them today. Noticeably, it also had the unfortunate side effect of creating a schism between the sciences and the humanities, a schism that still persists.

The Russian Nikolai Kondratev (1892 to 1938) observed that long business cycles, known as Kondratev's waves, have been prevalent since the beginning of the Industrial Revolution. Later Joseph Schumpeter (1883 to 1950), economist and sociologist, elaborated on this theme and founded modern economic growth theory. According to Nicholas Valéry, he challenged classical economics *"as it sought (and still seeks) to optimize existing resources within a stable environment – treating any disruption as an external force on a par with plagues, politics and weather"* [114] and associated the phenomenon of *"creative destruction"* with Kondratev cycles. Valéry maintains that *"as Schumpeter saw it, a normal, healthy economy was not one in equilibrium, but one that was constantly being "disrupted" by technological innovation"*.* He has extended Schumpeter's long economic waves to the present day, and concludes that the waves are shortening, from 50 to 60 years to around 30 to 40 years. In fact, Schumpeter's waves of innovation portray a fine structure on the carbocultural steps of the Staircase of Socio-Technological Development (see Figures 6.5 and 6.6). Though the waves are rather different, they all thrive on flows of heat from burning fossil fuels, including the digital wave's virtual world-wide web, which sometimes seems purely immaterial.

The Carbocultural Regime unleashed an explosion of cultural diversity, but in the turmoil new uniformity arose. Think of nation states and computer networks as examples. Furthermore, despite the fundamental differences between economies in the world, all accept money for the exchange of goods and services, and money is exchangeable itself.

---

*Note that in energetics terms economies in *"equilibrium"* are 'dead' economies; what economists mean is 'steady state', which is just far from equilibrium; every 'economic state' is sustained by structuring flows of energy.

Energy – Engine of Evolution

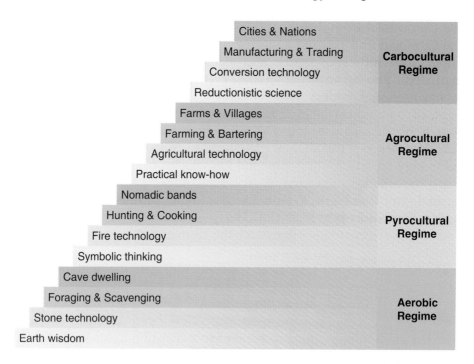

Figure 6.5   Staircase of Socio-Technological Development

The key words that attempt to characterise the anthroposystems of the Carbocultural Regime are *reductionistic science* for human knowing, *conversion technology* for human capacity, *manufacturing and trading* for human acting and *cities and nations* for the way people are living together.

The Staircase of Socio-Technological Development models the cultural evolution of humanity both in terms of the dominating energy regimes and as co-evolutionary processes of *human knowing, human capacity, human acting* and *human living*. The model does not imply that with every

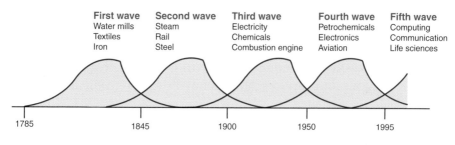

Figure 6.6   Schumpeter's waves of innovation

The Staircase of Energy Regimes

socio-technological revolution the past is wiped out. Just the contrary is true. The rhythms of socio-technological development keep resounding through time, which renders the evolution of human culture an energy-driven, polyphonic composition (see Table 6.1).

The pushes and pulls of progress

At each interface between the steps of the Staircase of Socio-Technological Development two similar *processes of progress* can be distinguished, a *push* and a *pull* mechanism. Pushes autonomously create needs in a step above, but at each step a need can emerge which induces a push from a step below; thus a pull actually is an induced push. So considered the Staircase invites us to extend the well-known concepts of *technology push* and *market pull* to the interplays of its four steps, as shown in Figure 6.7. The human needs corresponding to human knowing, human capacity, human acting and human living – or the four steps of the Staircase of Socio-Technological Development – are intellectual need, technical need, customer need and societal need, respectively. Note that whereas pushes can escalate upwards, pulls can cascade downwards: for example, a market pull can trigger a technology pull.

History shows that during cultural evolution the relative importance of human needs can, and do change. In his book 'Insight in Innovation: Managing innovation by understanding the Laws of Innovation', Jan Verloop explains how during the Industrial Age innovation in companies evolved from curiosity-driven to opportunity-driven, and from a cascading approach – from science to technology to business to society – to a bridge-building model centred around innovative combinations of technologies and markets [111b]. In terms of the processes of progress described here,

Table 6.1 The symphony of evolutionary energetics

	Aeroculture	Pyroculture	Agroculture	Carboculture
Human knowing	Earth wisdom	Symbolic thinking	Practical know how	Reductionistic science
Human capacity	Stone technology	Fire technology	Agricultural technology	Conversion technology
Human acting	Foraging and scavenging	Hunting and cooking	Farming and bartering	Manufacturing and trading
Human living	Cave dwelling	Nomadic bands	Farms and villages	Cities and nations

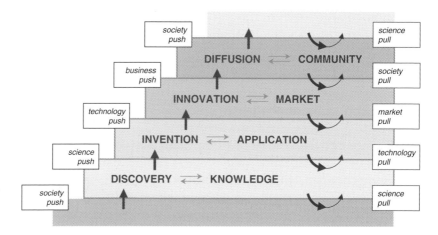

Figure 6.7 Pushes and pulls in socio-technological development

that change is consistent with a shift from autonomous pushes from science and technology to a system based on the duality of market pull and technology push.

THE EVOLUTION OF SYSTEM EARTH

The Staircase of Energy Regimes now stands complete and is ready to be applied in exploring pathways to the future. By zooming in, the Staircase model allows us to observe and discover patterns in human history. These patterns should help us to 'look' into the future, stimulating purposeful exploration for sustainable solutions. Some of the key patterns uncovered are summarised below.

The Staircase of Socio-Technological Development suggests that behind the scenes of a socio-technological revolution in effect six revolutions operate in concert. A socio-technological revolution, such as the Pyrocultural, the Agricultural or the Carbocultural Revolution, is *driven* by an energy revolution at its core, and *shaped* by the 'co-revolution' in human knowing, human capacity, human acting and human living, induced by a revolution in consciousness. Both the pyrocultural Symbolisational Signal and the agricultural Quantificational Signal fertilised a new perception of reality with Man, and each 'cultural signal' was a trigger for change.

The concept of co-evolution is key. It occurs at the scale of organisms, where flowers use bees for reproduction, giving them nectar for their services [28f]; it occurs in the ecosphere, which maintains closed loops through uncountable feedbacks; and it occurs in the anthroposphere,

The Staircase of Energy Regimes

where memes shape minds, minds shape artefacts, artefacts shape memes, and memes plus artefacts shape culture. But co-evolution also occurs at the global scale, where ecosphere and anthroposphere evolve together in moulding 'system Earth'.

The Staircase of Socio-Technological Development not only depicts the evolution of the anthroposphere as a whole, but also of its individual composing anthroposystems. Human acting, for example, evolved from foraging and scavenging, to hunting and cooking, to farming and bartering, to manufacturing and trading in the form of money-based exchange of goods and services. Thus the economy evolves as a cultural construct shaped by memes and artefacts – the only laws being immutable are nature's.

Maybe the most intriguing learning is that almost everything that lives does so at the mercy of a sunborne energy flow.* As a living 'something' is always an energy-dissipating structure, energy is the engine of evolution – the nature of which is change. Therefore the notion of 'sustainable development' actually means 'sustainable change'. It is this apparent contradiction in terms that defines the topic of the final chapter.

---

*Except from nuclear and geothermal energy.

– 7 –
The Emerging Helio-Energy Revolution

SIGNS OF A COMING ENERGY REVOLUTION

The Macroscopical Signal

The Pyro-Energy Revolution marked the origin of human civilisation. Its corresponding first human energy culture, pyroculture, was evolutionarily most successful and established ecological dominance over the animal world – although in the end it turned against itself and began to overshoot the carrying capacity of its habitat. However, a newly adopted perception of reality – as revealed by the Symbolisational Signal – allowed *Homo sapiens* to reinvent his societal metabolism. He increased the carrying capacity of the Earth for his own population by unleashing an energy revolution, the Agro-Energy Revolution. The new energy culture, agroculture, was evolutionarily most successful and established ecological dominance over pyroculture. However, suddenly, agriculture turned against itself and began to overshoot the carrying capacity of its habitat. A newly adopted perception of reality – exposed by the Quantificational Signal – allowed *Homo sapiens* to reinvent his societal metabolism. Again, he increased the carrying capacity of the Earth for his own population by launching an energy revolution, the Carbo-Energy Revolution. The third human energy culture, carboculture, is evolutionarily most successful and has established ecological dominance over agroculture in record time. Yet history repeats itself. The currently dominant energy culture is turning against itself by overshooting the carrying capacity of its habitat.

The evolutionary march of genus *Homo* proceeds in surges: after a gradual but significant rise all through the

Helian Emerging	
Carbian About 400 years ago	
Agrian About 12,000 years ago	
Pyrian About 0.5 million years ago	
Oxian About 2.1 billion years ago	
Photian About 3.8 billion years ago	
Thermian About 4.2 billion years ago	

Pyrian and the Agrian periods, the human footprint* started to develop explosively during the Carbian. As a consequence, say some biologists, the ecosphere perhaps approaches *"the sixth major extinction event in its history"* [118]. It must be emphasised that such future projections are highly uncertain [120a], but, if it were to happen, the Carbian Explosion would look like the impact of a giant asteroid ... yet in slow-motion. In any case, it seems that for the first time in history humankind is on a collision course with biophysical reality on a global scale [117]. It is this very notion that makes the current carbocultural overshoot essentially different from the earlier pyrocultural and agrocultural overshoots. Now *Homo sapiens* can no longer enlarge his habitat. This time he has no choice other than to increase the carrying capacity of the Earth by reducing his ecological footprint. According to the historic patterns depicted by the Staircase model, a radical increase in the carrying capacity necessitates a next energy revolution and corresponding socio-technological reform. That, in turn, would require the adoption of a new perception of reality, and thus a new 'cultural signal'.

It is not too long ago, in 1968, that the Apollo 8 crew focused a television camera on planet Earth [121]:

"For the first time humanity saw its home from afar, a fragile "blue marble" hanging in the blackness of space."

Since then 'macroscopical' observations using satellite monitoring have advanced rapidly [19b]. Generally, a 'macroscope' (from *macro*, great, and *skopein*, to observe) is *"a symbolic instrument"* composed of scientific methods and modern technologies borrowed from different disciplines. Joël de Rosnay 'invented' the tool which is aimed at studying the infinitely complex [122]. For example, these days a 'macroscopical thermometer' measures the global average temperature on that *"blue marble"*.

---

*The so-called ecological footprint is *"the area of productive land and water that people need to support their consumption and to dispose of waste"* [115]. In other words, it quantifies the total area required to produce the food and fibres that a country consumes, to sustain its energy consumption, and to give space for its infrastructure [116]. According to the World Wide Fund For Nature the ecological footprint grew by 80 percent between 1961 and 1999 to a level 20 percent above the Earth's biological capacity. Some are skeptical about the concept, like, of course, Bjørn Lomborg, but many researchers see the measure as useful [115]. They argue *"footprints help to summarize information, and are no more of a gimmick than such well-accepted yardsticks as gross domestic product"*. Says Simon Levin: *"No single indicator is a good measure, as every time you collapse lots of diverse information you lose something."* I agree with William Rees who originally proposed the concept, that the ecological footprint brings in the notion of 'carrying capacity' [117], which, in my view, makes the key difference between conventional health, safety and environmental (HSE) policies and sustainable development.

In addition, scientists are reconstructing the temperature course of the planet's history from macroscopical paleo-data. In effect, advanced macroscopical Earth-observation systems enable humankind to *"watch its own footprint"* on the planet [123], and to observe the processes of change on a global scale, even retroactively. More and more economical, ecological and social indicators, from the stock market price indices to the WWF's Living Planet Index or the UN Human Development Index, comprise macroscopical information. All together they create a new cultural signal, the 'Macroscopical Signal'. Now, through 'zooming out' from 'system Earth' with macroscopes, instead of zooming in on matter with microscopes, or on celestial bodies with telescopes, Homo sapiens is beginning to discover how natural and cultural processes interrelate.

The Macroscopical Signal is an immense historic achievement for the carboculture (in terms of evolutionary energetics it is an 'emergent property' of the carbocultural energy-dissipating structure). After making his immediate vicinity debatable through speaking symbolically, and then making his environment intelligible through measuring quantitatively, at this time Homo sapiens is making his footprint on the globe observable through researching macroscopically. However, observing is one thing, interpreting is another, let alone taking subsequent action. This is because the nature of the response depends critically on people's perception of reality.

Two perceptions of reality

Historian Donald Worster argues that from the Age of Reason *"a most discordant set of ideas"* emerged in the form of two major traditions [124a]:

> *"The first was an "arcadian" stance toward nature ... which advocated a simple, humble life for man with the aim of restoring him to a peaceful coexistence with other organisms. The second, an "imperial" tradition ... [aspired] to establish, through the exercise of reason and by hard work, man's dominion over nature."*

According to Worster, Francis Bacon (1561 to 1626), the influential enlightened philosopher, was an early proponent of the imperial view [124b]:

> *"Bacon promised to the world a manmade paradise, to be rendered astonishingly fertile by science and human management. In that utopia, he predicted, man would recover a place of dignity and order, as well as authority over all the other creatures he once enjoyed in the Garden of Eden. Where the arcadian naturalist exemplified a life of quiet reverence before the natural world, Bacon's hero was a man of 'Active Science', busy studying how*

he might remake nature and improve the human estate. Instead of humility, Bacon was all for self-assertiveness: 'the enlargement of the bounds of Human Empire, to the effecting of all things possible'.... 'The world is made for man', he announced, 'not man for the world'.... Science offered the means for building a better sheepfold and creating greener pastures."

By and large, Western civilisation followed the direction Bacon envisaged in his work 'The New Atlantis' [125a], particularly when we look at the technological fruits of the scientific method.

Entirely in line with his historic roots, *Homo ecologicus* tries *"to dispel modern society's confidence in technology, and more, its faith in unlimited economic growth"* [124c]. He attacks the focus of *Homo economicus* on continuous economic growth, which does not yield sustainable welfare, but depletion of natural and social capital instead [126a]. The title of Richard Douthwaite's book almost tells the story on its own: 'The Growth Illu$ion: How economic growth has enriched the few, impoverished the many and endangered the planet' [127]. That *"mad dream of humanity"*, adds de Rosnay [122], more subtly paraphrased by Gerald Marten [128a]:

"Modern values about material possessions are connected to our perception that economic growth is essential for a good life. Political leaders tell us that economic growth is their highest priority, while "experts" addressing us through the mass media continually reinforce our belief as a society that a high level of consumption (consumer confidence) is essential for full employment and a healthy economy ... [But] continual expansion of material consumption is ecologically impossible."

In our time, says Worster, *Homo ecologicus "has come to represent the arcadian mood that would return man to a garden of natural peace and piety"* [124c].

Physicist Michio Kaku, in accord with the imperial vein, paints the 'Next Atlantis' after Bacon's 'New Atlantis' in his 'Visions: How science will revolutionize the 21st century'. Kaku sees humanity standing on the cusp of an epoch-making transition from passive observer to active choreographer of nature [129a]. *Homo sapiens* may well be creating the bodies for the next stage of human evolution [129b], *Robo sapiens*, a blend of human and mechanical properties with superior survival possibilities [130a]. Instead of watching the dance of life, genetic modification *"will ultimately give us the nearly god-like ability to manipulate life almost at will"* [129c]. Already bio-scientists are exploring concepts such as 'Gene Doping' [131] and 'Synthetic Life' [132]:

"Biologists are crafting libraries of interchangeable DNA parts and assembling them inside microbes to create programmable, living machines ...

The Emerging Helio-Energy Revolution

Evolution is a wellspring of creativity ... But there is still plenty of room for improvement."

Kaku believes that Man will make the transition from unravelling the secrets of nature to becoming master of nature, with quantum mechanics, biogenetics and artificial intelligence providing the control sticks [129c]; Lomborg adds *"geo-engineering"* of the whole ecosphere to the dashboard [133a].

Leaving Carbon Valley

In line with their antipodal perceptions of reality, or worldviews, the two cultural subspecies of *Homo sapiens* – Imperial Man and Arcadian Man – show diametrically opposed responses to the Macroscopical Signal. Although, remarkably, both plan to leave Carbon Valley, the land ruled by the Carbocultural Regime, to head for a more sustainable future.

No doubt conventional oil and gas stocks will decline this century, with oil preceding gas. Still, king coal has the potential to do well for a couple of hundreds of years; oil and gas could pass the next turn of the century provided unconventional sources* are produced to the full; and when gas hydrates[†] are taken into account, hundreds, or maybe thousands, of carbocultural years potentially remain. Yet, as the pyroculture did not disappear because of a lack of firewood, and the agroculture did not disappear because of a lack of sunlight, the carboculture will not end

---

*The natural environment accommodates enormous resources of unconventional oils *"amounting globally to more than 400 Gtoe, with about two-thirds of that in oil shales, roughly one-fifth in heavy oils and the rest in tar sands"* [134b]. Theoretically, non-conventional oil resources could lengthen the oil age by more than a century at current supply rate. Non-conventional gas embraces resources that are already being recovered such as methane in coal-beds as well as large deposits in tight reservoirs and high-pressure aquifers, in addition to the methane hydrates discussed above for which appropriate production technologies are lacking [134c].

[†]It is estimated that the global mass of organic carbon locked in gas hydrates is *"roughly twice as much as the element's total in all fossil fuels ... Tapping just one percent of the resource would yield more methane than is currently stored in the known reserves of natural gas."* [134c] In gas hydrates methane is *"trapped inside rigid lattice cages formed by frozen water molecules"*. These ice-like compounds can stay close to 100 m deep in continental polar regions, about 300 m in oceanic sediments, and about 2000 m in warmer oceans. Winning the often quite diluted gas molecules without losing them to the environment is a devilish job: methane has a global warming potential that is 21 times that of carbon dioxide [134d]. Says Vaclav Smil [134c]: *"Not surprisingly, pessimists consider hydrate recovery to be an illusion, while many enthusiasts are thinking about plausible means of extraction."*

because of a lack of fossil carbon. Says Vaclav Smil, there is *"a very low probability"* that fossil fuels will energise the world over the next hundred years to the same extent as they do now [134a]:

> *"The intervening decline of our reliance on coal and hydrocarbons and the rise of nonfossil energies will not take place because of the physical exhaustion of accessible fossil fuels but ... because of the mounting cost of their extraction and, even more importantly, because of environmental consequences of their combustion."*

History teaches that the sustainability of an energy regime and the availability of its prime energy resource are two different things.

Both the Imperials and the Arcadians envision a future with a fundamentally different energy regime, and, correspondingly, with an entirely different societal metabolism. But of course their images of a sustainable future are poles apart: while Imperial Man explores pathways to Nuclear Valley, Arcadian Man wants to head towards Green Valley.

Nuclear Valley

Since Imperial Man dominates today's Carbocultural Regime, the energy policies of the developed world give a clear indication of his ideas. Nowadays the fossil–fuel and nuclear industries *"get the lion's share of subsidies"* [135a]. Moreover, the allocation of research and development (R&D) budgets shows the imperial preference for nuclear energies with regard to the future. For many decades, members of the International Energy Agency (IEA) have been spending a substantial share of their total R&D budget on nuclear fission. Further, between 1974 and 2001, the IEA states have spent more money on nuclear *fusion* R&D than on all renewable energy systems together. In 2001, the R&D budget ratio of nuclear (fission plus fusion) to total renewables was roughly 6 : 1 [136a].*

Like all anthropogenic technologies, nuclear technologies could evolve along a trajectory of advancing 'steady states' in a co-evolutionary play between human ingenuity and technological capacity – or between memes and artefacts. At the end of 2003, the first of a so-called 'third-generation reactor' was ordered, with improved safety and productivity parameters compared to the state-of-the-art 'second generation' reactors which date back to the 1960s [137a], and whose combined capacity now produces over 16 percent of the world's electricity [134e]. Whereas the

---

*The technology fields are conservation, fossil fuels, renewable energy, nuclear fission, nuclear fusion, power and storage, and other.

third generation, conceived a decade ago, is actually a continuation of existing technologies [137a], researchers have now set their sights on a radically new fourth generation. Within the next 15 to 20 years, innovative concepts should lead to the better use of nuclear fuel and reduced nuclear waste, while simultaneously offering further increased safety and lower production costs.

But is there enough nuclear fuel to establish a Nucleocultural Regime? Economists say 'no', stating that nuclear reserves will last for only another half century [138], whereas physicists say 'yes', claiming that humankind can utilise fission fuel for many centuries to come, or even for millennia. The economists have present rates of uranium consumption in mind and contemporary conversion technologies, whereas the physicists think about the next generations, about adding thorium to the nuclear fuel (which is not only much more plentiful – about five-fold – but also more accessible than uranium [134f]) and, last but not least, about breeding that potentially provides *"virtually infinite energy reserves"* [138].

However, despite the great promise of nuclear *fission*, in fact nuclear *fusion* would be the crowning glory of Imperial Man's inventiveness. And nuclear *fusion* is a totally different story than nuclear *fission*. After decades of research, fusion has been demonstrated for a time span of about one second, the main obstacle being a required temperature of 50 million degrees Celsius [139]. Recently, the international fusion community agreed the design of the International Thermonuclear Experimental Reactor (ITER) aimed at demonstrating the technical feasibility of electrical power production from nuclear fusion. The construction of the ITER will take about 10 years. Then, after 10 to 20 years it must become clear whether a pilot power plant can be brought on line approximately 35 years from now [140]. This, in turn, could pave the way to the first commercial fusion power plant towards the middle of this century.

The fusion reaction converts the two hydrogen isotopes, deuterium and tritium, into helium [140]. Tritium must be produced in the fusion reactor from lithium, an element abundantly present in the Earth's crust and seas. From the mining seams of lithium, theoretically, the world could be energised for thousands of years, while lithium extraction from oceans would push back this limit to several million years (according to the French Atomic Energy Commission*). Also, deuterium, the other fusion fuel needed, can be produced from ordinary seawater. The vastness of its resources surpasses all imagination, representing more than 10 billion years of average annual world energy consumption. Therefore, it is

---

*See Clefs CEA report No. 49, Spring 2004 (French Atomic Energy Commission).

Imperial Man's greatest wish to fuse two deuterium isotopes instead of a deuterium and a tritium nucleus. Deuterium-deuterium fusion, which has been demonstrated recently in the laboratory [139], would yield a practically unlimited flow of energy … a sort of 'man-made sun' on Earth.

In the eyes of Imperial Man, developing nuclear power is out-and-out sustainable. It secures energy supply, and thus economic growth, and therefore the advancement of living conditions for all people. Notwithstanding the improvements in fission technologies envisioned, Imperial Man assigns a sustainability advantage to nuclear fusion. Fusion is intrinsically safer than fission, as a failure or uncontrolled operation immediately leads to reactor shut down [140]. Also, fusion fuels are not subjected to non-proliferation treaties. And fusion is cleaner than fission: both operate without emitting greenhouse gases, but unlike a fission plant, fusion produces no long-lasting radioactive decay products. Fusion does make the reactor structure radioactive, but Imperial Man expects to mitigate this by using new construction materials [139].

As well as nuclear fuels and nuclear technologies, a nucleocultural societal metabolism must include (an) energy carrier(s) to provide people with the energy where it is needed in Nuclear Valley. Nuclear reactors, both the fission and the fusion type, produce heat. This heat then is converted – just as it is with today's carbocultural coal- and gas-fired power plants – into electricity, which in turn is transmitted and distributed, through large-scale electricity grids, to its final use in stationary applications. With electricity being the most convenient energy carrier, you might think an all-electric Nuclear Valley would be the preferred future image of Imperial Man. But that is not necessarily the case as he doubts the aptness of electricity for modern mobility, perhaps one of the greatest achievements of carboculture. Therefore, as a second energy carrier, hydrogen plays a central role in Imperial Man's nuclear vision. Hydrogen could be made from water through electrolysis with nuclear electricity. When brought into a fuel cell, hydrogen recombines with airborne oxygen, yielding electricity on demand while emitting warm water.

Imperial Man foresees that journeying along the road from Carbon Valley to Nuclear Valley will take time. Therefore, to sustain socio-technological progress, he needs to advance the fossil fuel economy in the meantime by developing non-conventional oil and gas resources, and reviving coal. But 'new fossil' should be 'clean fossil', or fossil fuel usage that does not lead to emissions. The greenhouse gas carbon dioxide then must be sequestered. Several routes are being explored, from forest and soil management, to injection in geological formations such as underground aquifers and deep oceans, and through chemical fixation [141]. Because carbon dioxide sequestration necessitates capturing carbon dioxide, and efficient carbon dioxide capture goes with hydrogen production, Imperial Man paves the way to 'nuclear hydrogen' via 'fossil hydrogen'.

Let me conclude with a few words about the imperial view of socio-technological development in general. Imperial Man believes in the fortunes of the free market economy leading to ever increasing prosperity through its 'invisible hand'. He addresses global issues through global partnerships, if only to smooth the globalisation of the free market. In his eyes, humankind is on the right track, perhaps slowly, but steadily. For Imperial Man, sustainable development is above all a grand techno-scientific challenge. He expects a lot from nanoworks, including its bio- and digital incarnations, when it comes to better controlling life and the environment. Francis Bacon who *"promised to the world a manmade paradise, to be rendered astonishingly fertile by science and human management"* keeps with him.

Green Valley

Where Imperial Man chooses Nuclear Valley as a destination, Arcadian Man steers towards Green Valley. Green Valley has no nano-works, as nano-works aim at artificial life and thus threaten Nature. Likewise, in Green Valley genetic engineering is entirely banned. Arcadian Man's no-growth economy aims just to conserve Nature, as only then flora, fauna and 'humana' have a chance to live together harmoniously and peacefully. Green growth will be above all spiritual, and not material. Being a true egalitarian, Arcadian Man

- shares the planet with all other biological species;
- treats all individuals equitably so as to eliminate the rich-poor gap, and;
- takes into account both current and future generations [119].

Arcadian Man's vision may include the characteristic voices of 'civil society', the rather diverse whole of non-governmental organisations and allied academics. For example, an Arcadian stance with regard to nuclear energy comes from Greenpeace. Under the title 'Europe Continues to Back Nuclear Pipe Dream' a telling press message, 26 November 2003, reads:

> *"Pursuing nuclear fusion and the ITER project is madness ... Nuclear fusion has all the problems of nuclear power, including producing nuclear waste and the risks of a nuclear accident. Why is Europe backing a bad energy option, with no prospect of operation in the near future, when alternative, environmentally acceptable options for electricity generation exist now? Renewable energy has massive potential, yet the EU continues to plough billions of euros in research and development grants into nuclear fusion."*

Arcadian Man views the idea of nuclear fusion as pure megalomania.

A green societal metabolism feeds on solar energy and cycles matter in closed loops like Nature. Only then can anthropogenic disruptions of the ecosphere remain small enough to prevent its degradation, and thus the coupled corrosion of mankind's own economy which he views as a pure subsystem of the ecosphere. Being a technology pessimist, Arcadian Man does not believe that he will ever be able to create closed material loops with current – read: imperial – consumption fluxes. That is why in his future image of Green Valley an Ecocultural Regime rules an economical – in the sense of frugal – dematerialised, highly resource efficient, solar-driven economy.

The planet's global intercept of solar radiation amounts to roughly 170,000 TW (1 TW = 1000 GW) [33q]. Through all photosynthesisers, Nature consumes 100 TW. The anthropogenic energy flow is about 14 TW, of which fossil fuels constitute approximately 80 percent [142]. Future projections indicate a possible tripling of the total energy demand by 2050 [143], which would correspond to an anthropogenic energy flow of around 40 TW. Of course, based on the Earth's solar energy budget such a figure hardly catches the eye, but in comparison to the biosphere it is really sizable. And in Arcadian eyes the Western energy throughput already today *"clearly poses a threat to eco-systems world-wide"*.

But what then is the Arcadian limit? In 'Sharing the Planet: Population – Consumption – Species' Johan van Klinken notes an *"admissible total human energy consumption"* of 260 EJ per year,* which equals a human-induced energy flow of 8.2 TW [95a]. For a projected world population of about 9 billion people in 2050,† this 8.2 TW limit yields an energy budget of 900 W for each person. In another contribution to 'Sharing the Planet', Lucas Reijnders states that rapidly advancing applications for the sustainable harvesting of renewable energy fluxes (solar, wind and geothermal) have the technical potential to fully replace fossil fuels *"as they are currently used"* [144a]. That boils down to an energy budget of 1.2 kW per world citizen in a 9 billion people world.‡ By and large, Green Valley will be 'a 1 kW society'. To put this number in perspective, today a European consumes on average 6 kW energy, while an American uses about 12 kW [95a].

---

*Johan van Klinken quotes Dürr and Ziegler who both arrived tentatively at the same limit.

†Projections recently issued by the United Nations suggest that world population by 2050 could reach 8.9 billion, but in alternative scenarios could be as high as 10.6 billion or as low as 7.4 billion [148].

‡In 2002, the total fossil primary energy supply was 8143 Mtoe [142].

It seems that Arcadian Man must make great sacrifices to reach Green Valley, but nothing is further from the truth. In the first place austerity wards off materialism, which only alienates him from Nature and his fellow man. Also, reducing energy consumption by a factor of 6 to 12 does not look too bad to him, because resource efficiencies can be boosted by a factor of four or more [145,146] with already existing or almost-existing technologies. However, the Arcadian bottom line is that we are talking about a Western luxury issue here, since for many people in the world a 1 kW energy allocation would improve the quality of life significantly.

Arcadian Man's fuel preferences are electricity and biofuels. In Green Valley, electricity comes solely from modern renewables such as solar, wind, small hydro-, wave and tidal power. Besides, in Green Valley operate small-scale biopower and biofuel manufacturing units, especially in biomass-rich environments like rural areas, and in wood and food production fields. While solar panels and windmills deliver their electricity intermittently with the coming and going of sunlight and wind, green plants store solar energy in carbohydrates. Of course, the Arcadian energy system does not interfere with the food economy, but just cascades with it through the conversion of its energy-rich, and thus valuable, residues. Arcadian Man is sceptical about hydrogen as it is an energy carrier and not an energy source. Perhaps hydrogen makes an energy source more convenient to use, but its production, just like its storage and distribution, is potentially capital and resource intensive and thus could, one way or another, create too large a footprint.

In all he does Arcadian Man follows the motto 'Small is Beautiful' [147]. He wants his technologies to fit human dimensions so as to keep them manageable and to prevent unplanned, maybe uncontrollable side effects. But also, 'small' benefits from economies of numbers. This enables distributed generation and manufacturing, which improves the resilience of sociosystems because 'small' reduces risks: the bigger the technological artefact, the bigger the calamity in the event of a natural disaster, an accident or human sabotage. Moreover, 'small' unites people with their environment, which increases awareness of ecosystem services, and 'eco-awareness' is the enabler of true innovation. Further, 'small' employs people in the region, and with employment comes prosperity, satisfaction, engagement, and thus community building, which in turn boosts societal resilience. Small is unobtrusive, too, preserving the greenness of the valley. And, last but not least, 'small' is affordable, which enables socio-economic development in the world's less developed regions, including rural areas. This in turn restrains detrimental urbanisation.

Arcadian Man's metabolic pathways to Green Valley look like these: stop exploring for new coal and oil reserves, as the Earth cannot even assimilate the carbon dioxide emissions resulting from burning the current stocks; switch from carbon-rich coal to carbon-poor natural gas in power generation while successively expanding green electricity supply; rigorously improve resource efficiencies by introducing efficient end-use appliances, closed-loop technologies, lifestyle changes, and demand side management aimed at about 1 kW for every human being on Earth; boost the production of modern renewable energy and recycling solutions as per today, thus learn by doing; and develop the solar-electricity and biomass-to-biofuels economies with (almost) existing technologies. But above all sustainable development remains a socio-ethical and a corresponding socio-economical challenge to Arcadian Man. It is about equity and about eradicating poverty in the first place.

Sustainable development

In line with their opposing perceptions of reality, the Imperial and Arcadian responses to the Macroscopical Signal differ sharply, reflecting opposing interpretations of the notion of sustainable development. I would like to introduce some key sustainable development criteria from an evolutionary energetics point of view, and then evaluate the Imperial and Arcadian perspectives against them.

As we have seen in the previous chapter, 'system Earth' can be viewed as an energy-dissipating superstructure built from innumerable co-evolving eco- and sociosystems, which are all energy-dissipating structures in themselves. None of them functions in sole isolation: they are all, more or less, coupled to each other through uncountable and largely unknown stabilising and destabilising feedback loops. Their life-giving energy flows, again and again, induce the emergence of new properties, or manifestations of wholes being more than the sum of their parts, through synchronised interactions, the 'invisible hands'. Because each eco- or sociosystem evolves far from thermodynamic equilibrium, all show unpredictable threshold behaviour: when being forced, a system's tendency to maintain its structure can be broken; then it adopts a new structure with a more complex metabolism, utilising more energy and matter, or it either degrades or collapses.

As well as natural forcing, also anthropogenic forcing can push an energy-dissipating structure such as an eco- or a sociosystem out of its dynamic stability domain. Since earthly energy-dissipating structures are complexly interconnected, such a change will inevitably be 'felt' by other living systems: if one energy-dissipating structure cannot counteract certain anthropogenic forcing, and degrades or collapses, it could drag along others, including the forcing sociosystem itself. Of course, an energy-dissipating structure's capacity to counteract external forcing through adaptation differs for each system. The better a dissipative structure performs in this respect, the more resilient it is; resilience is the ability to balance stability and adaptability simultaneously, two easily opposing requirements. Nonetheless, the notion of resilience brings us to the core of sustainable development from an energetics angle: sustainable development is socio-economic development that promotes the resilience of eco- and sociosystems through controlling anthropogenic forcing. But what does this mean, practically?

Macroscopic observations reveal that the currently dominating Carbocultural Regime emits massive amounts of anthropogenic residues into the environment [98b]:

> "50 to 90 percent of the mass of industrialised-country environment outflows goes up into the atmosphere ... Waste products of economic activities are, for the most part, increasing ... Today's economies act as a linear system: materials and energy are taken from the natural environment, put to a brief useful life, and then become waste in the atmosphere, on land, or in water."

These words do not come from some radical Arcadian source, but from 'Tomorrow's Markets: Global Trends and Their Implications for Business', a co-production of the World Resources Institute, the United Nations Environment Programme and the World Business Council for Sustainable Development. The carbocultural socio-metabolism boils down to the predominantly linear, once-through conversion of natural resources into human waste. And once-through conversion inevitably depletes natural resources while burdening the environment. What is more, anthropogenic flows of economic inputs and rejected outputs have reached global dimensions, which was different when the first coal-fired industries heralded the Carbian Period about 400 years ago. Then, for example, the anthropogenic carbon dioxide emissions could be assimilated by green plants. But today humanity emits more than 24 Gtons of this greenhouse gas per year [142]. And only a part can be processed by the photosynthetic troops, as the rising carbon dioxide concentrations in ocean and air proof. Within a few centuries the relatively harmless human-induced carbon dioxide emission has turned into a global force of change. *Homo sapiens carbonius* has substituted a global economy for a niche activity without adapting the fundamentals of his socio-metabolism. To reduce the resulting global anthropogenic forcing leaves him with no other option – apart from going back to niche-life again – than transforming his linear, once-through metabolism into a cyclic, near-closed-loop metabolism, like Nature's. Only a closed-loop* metabolism can keep the depletion of natural resources and its coupled burdening of environmental sinks within the Earth's biophysical limits.

Obviously, in a closed-loop economy physical matter cycles, but what would be the requirements for energy? Inevitably, a closed-loop metabolism needs to feed on a mass-less energy flow, such as heat or light – otherwise it consumes a resource and loads a sink in a once-through conversion. A relatively small flow of mass-less energy springs from the Earth's hot inner sphere in the form of geothermal heat, but a powerful flow of mass-less energy reaches the globe's surface in the form of sunlight. As a corollary, since only primary producers such as photosynthesisers and photovoltaic cells can harvest solar energy directly, secondary producers such as mammals and manufacturing industries need to be predominantly fuelled by 'solar-charged' energy carriers.

I note that in view of the energetic evolution of life, the sustainable energy criterion 'renewability' seems far from strange: all energy regimes, up to the last one, were driven by incident sunlight and renewable energy carriers. Even *Homo sapiens carbonius* cannot live without fresh solar energy: in the first place, he is a heterotroph, which means that he depends on prime producers like green plants for the provision of the biochemical building blocks his body needs for bio-maintenance; in the second place, he is an aerobe, which means that he needs natural oxygen for the *in vivo*, cellular combustion of carbohydrates; in the third place, he is a solar farmer, which means that he needs sunlight to produce more food than the land provides naturally; and in the fourth place he is a fire master, which means that he needs oxygen from green plants for the combustion of fossil fuels.

Still, today's energy challenge is unparalleled in history: the phenomenal 'ancient sunlight' stocks set free by carboculture must be compensated with incident sunlight to further socio-technological development, but sustainably. From an evolutionary energetics angle, the crux to sustainable development lies in creating 'resilient socio-metabolisms' based on 'recyclable matter' and 'renewable energy' – or recyclables and renewables – with a view to minimising anthropogenic forcing on ecosystems so as not to affect their resilience.

---

*It would be correct to write 'near-closed-loop metabolism'. Also Nature does not operate in a perfect closed loop mode – think, for example, of the ancient fossil remains fuelling present carbocultural economies. Yet just to be concise I write 'closed loop'.

The Emerging Helio-Energy Revolution

Limits to nuclear

Imperial Man heads for Nuclear Valley along the pathways of clean fossil fuel economies. But neither nuclear fission nor nuclear fusion fuels are renewable. Hence nuclear conversions deplete natural resources and leave anthropogenic residues, albeit the differences between fission and fusion are large. Regarding fission, a sound solution for treating radioactive leftovers does not exist as yet [134g]. In addition, according to Mohamed ElBaradei, Director General of the International Atomic Energy Agency (IAEA), the nuclear fission economy's export control regime is failing – *"as evidenced by the recently discovered black market of nuclear material and equipment"* [149] – which poses a serious threat in the form of nuclear weapons proliferation. Further ElBaradei says, the risk of a severe nuclear accident *"can never be brought to zero"*. Add to this the danger of conflicts, the more so when fission expands, and also the notion that private companies insure against only a small part of the risks involved [150a]. Indeed, nuclear fission leads to hardly any greenhouse gas emissions and therefore helps to mitigate climate change, but safety and security issues seriously jeopardise the resilience of both eco- and sociosystems.

Imperial Man strongly supports the free market economy. But for some reason he sets his economic standards aside when it comes to nuclear power: large subsidy streams have flown and will flow to research and development without commercial achievements being required. Still, the huge allocation of public funds to nuclear as compared to renewables (about six to one today) echoes Imperial Man's neoclassical appreciation of the economic process [151a]:

> *"Neoclassical economists view natural and created capital as* substitutes *in production. They are technological optimists, believing that as resources become scarce, prices will rise, and human innovation will yield high-quality substitutes, lowering prices once again. More fundamentally, neoclassicals tend to view nature as highly resilient; pressures on ecosystems will lead to steady, predictable degradation, but no surprises. This view of small (marginal) changes and smooth substitution between inputs is at the heart of the broad neoclassical tradition in economics ... At the level of a broad vision, neoclassicals generally believe that the global spread of market-based economies provide a powerful foundation for achieving sustainable development."*

Do you hear, like me, the words of the founding genius of modern economics, Adam Smith, who *"saw nature as no more than a storehouse of raw materials for man's ingenuity"* [124d]? But Adam Smith lived in the

18th century, in a relatively small world. Today, in the middle of the Carbian Explosion, his statement no longer holds. The global economy touches the limits of once-through conversion of natural resources in anthropogenic waste, which *does* affect the resilience of Nature, as the Macroscopical Signal reveals. Yet Imperial Man's belief in the seemingly endless substitutability of natural resources – including matching environmental sinks – prevents him from switching to a closed-loop economy. His anthropocentric stance to Nature – all that Nature provides is being evaluated from the perspective of being useful or usable for economic development, up to the genes of biological species [152a] – as well as his reductionistic thinking – which renders typical energy-dissipating structure properties such as unpredictable threshold behaviour unthought-of – colours his judgement of macroscopical signals of change.

Imperial Man hardly realises how fast carbocultural evolution proceeds with respect to nature's evolution. For example, if climate change unfolds within a few decades climatologists speak about an *abrupt* alteration, but neoclassicals call an economic outlook of a similar period in time *long-term*. As a consequence, a neoclassically governed economy slowly adapts to rapid environmental change, if at all, which further corrupts its chances to sustain – for the arrow of time is inexorable. Though the issue is not the mere concept of the free market economy, as Arcadians argue, but the flawed translation of macroscopical signals of unsustainability into socio-economical preconditions limiting the 'development space' within the boundaries of sustainable development. For example, if disposing of the greenhouse gas carbon dioxide remains almost free of socio-economic constraints, then innovative carbon-dioxide-free energy solutions have no chance to compete. But an economic law is a cultural phenomenon. It can be revised consciously and purposely. Only the laws of Nature are immutable, such as the Laws of Thermodynamics. The latter reveal insight about uncertainty and irreversibility, but also about the creative power of energy-driven organisation. Adam Smith's 'invisible hand' can be guided – not controlled – through socio-economic governance.

Limits to green

Arcadian Man sets course to Green Valley via a short cut. He opts for renewable and recyclable solutions based on existing or novel technologies. What remains is the question whether the Arcadian route will yield resilient eco- and sociosystems. Just as with Imperial Man, in his perception of the economic process lies the key to understanding his view of a sustainable allocation of resources. In contrast to the neoclassical position [151a]:

> *"Ecological economists argue that natural and created capital are fundamentally complements – that is, they are used together in production and have low substitutability. Technological pessimists, ecologicals believe that as the sinks and the sources that make up our stock of natural capital are exhausted, human welfare will decline. Fundamentally, ecologicals view natural systems as rather fragile. If one component, say, of a fishery, is disturbed, the productivity of the entire ecosystem may plummet. This vision of web-like linkages between nature and economy has led this group to refer to itself as ecological economists. In contrast to neoclassicals, ecological economists consider the globalizing world economy to be on a fundamentally unsustainable path, and that further ecological pressure in the form of population and consumption growth is likely to lead to real disaster."*

Arcadian Man sees no other way of achieving closed material loops than to radically reduce energy and material flows. Remarkably, his rather pessimistic analysis appears to a large extent *"thermodynamically motivated"* [154a], thus based on the science of energetics, with reference to founding father Nicholas Georgescu-Roegen [154b]. A contribution by Stefan Baumgärtner to the Internet Encyclopedia of Ecological Economics entitled 'Entropy' reads [155]:

> *"In thermodynamic view, the economy uses low entropy energy and matter from its surrounding natural environment, to produce consumption goods, and discards high entropy wastes back into the environment. Georgescu-Roegen saw the Entropy Law as a metaphor for the inevitable decline of such a system, where every act of production and consumption brings the economy closer to doomsday in the form of 'heat death.'"*

In 1999, Georgescu-Roegen's reference work 'The Entropy Law and the Economic Process' was reissued by the prestigious Harvard University Press after its first publication in 1971. The back cover informs the potential reader:

> *"Every few generations a great seminal book comes along that challenges economic analysis and through its findings alters men's thinking and the course of societal change. This is such a book, yet it is more."*

And further:

> *"The entropic nature of the economic process, which degrades natural resources and pollutes the environment, constitutes the present danger.*

The earth is entropically winding down naturally, and economic advance is accelerating the process."

The author's underlying thermodynamic reasoning becomes apparent in the following passage [156a]:

"The flow of the sun's radiation will continue with the same intensity (practically) for a long time. For these reasons and because the low entropy received from the sun cannot be converted into matter in bulk, it is not the sun's finite stock of energy that sets a limit to how long human species may survive. Instead, it is the meagre stock of the earth's resources that constitutes the crucial scarcity."

But, all biological *"matter in bulk"* around us just emerged as a result of consuming *"the low entropy received from the sun"*! Since the origin of life, energy-dissipating structures have created gigantic amounts of *"matter in bulk"*, including the huge volumes of what we call fossil reserves. Indeed, humankind operates a once-through economy driven by low-entropy fossil fuels, but to conclude with the Second Law that human culture cannot achieve what nature has – that is operating in a near-closed-loop mode – is, thermodynamically speaking, not correct. Thermodynamics make no difference between biogenic and anthropogenic energy and material conversions. Georgescu-Roegen's theory is not thermodynamically based, but, as it seems, socio-ethically [156a]:

"Population pressure and technological progress bring ceteris paribus *the career of the human species nearer to its end only because both factors cause a speedier decumulation of its dowry ... we must not doubt that, man's nature being what it is, the destiny of the human species is to choose a truly great but brief, not a long and dull, career."*

And a little further on [156b]:

"Nor should we ignore the fact that the thorough reorganization of agriculture as proposed requires that a fantastic amount of resources now allocated to the production of durable consumer goods be reallocated to the production of mechanical buffaloes and artificial manure. This reallocation, in turn, demands that the town should abdicate its traditional economic privileges. In view of the basis of this privileges and of the unholy human nature, such an abdication is well-nigh impossible. The present biological spasm of the human species – for spasm it is – is bound to have an impact on our future political organization. The shooting wars and the political upheavals that have studded the globe with an appalling frequency during recent history are only the first political symptoms of this spasm."

If there is a way to sustain human civilisation, Georgescu-Roegen sees no other than a neo-agricultural way. His motto is 'back to the farm'.

Then a word about this term 'growth', economic growth, to be precise.

The ecological economist Herman Daly argues that only a so-called 'steady state economy' can be sustainable in the end, where 'steady state' appears a byword for no-growth: in a steady state economy *"the aggregate throughput is constant, though its allocation among competing uses vary free in response to the market"* [157a]. Similarly, and with reference to Georguescu-Roegen, Daly applies 'thermodynamic' reasoning:

> *"Finitude [as determined by the biophysical limits of the planet] would not be so limiting if everything could be recycled, but entropy prevents complete recycling."*

It is true that the economic conversion of lower-entropy natural resources leads to higher-entropy anthropogenic residues, but, if designed for, the higher-entropy residues can be converted into lower-entropy 'restarting' materials at the cost of lower-entropy sunlight – just as Nature converts higher-entropy water and carbon dioxide into lower-entropy carbohydrates. In Nature's economy carbon dioxide has become a 'recyclable', while in mankind's it is still a waste.

Undeniably, achieving a closed-loop economy is a phenomenal challenge, but there is no such thing as a thermodynamic limit to recycling. As always, the struggle for survival boils down to finding conversion pathways for energy and matter, or an evolutionary successful metabolic strategy. Nature's economy has grown phenomenally over billions of years, including severe setbacks, from small thermophilic microbial communities to the complex ecosphere we know today, while developing closed-loop processing. Of course, there are limits to 'once-through growth' for the trivial reason that once-through is not twice-through, but limits to 'closed-loop growth' are simply unknown. (I see no imminent limits to recycling frequencies within sociosystems.) One limit to economic growth is human ingenuity, the catalyst of economic development, and another the carrying capacity of planet Earth for the whole of eco- and socio-spherical energy-dissipating structures. However, Nature can proceed along its evolutionary course without humanity, while humanity cannot live without Nature.

Arcadian Man's ecocentric perception of reality – all that Nature provides has intrinsic value, like human life [152a] – as well as his holistic thinking – *"the universe and especially living nature is seen in terms of interacting wholes (as of living organisms) that are more than the mere sum of elementary particles"* [158] – shape his assessment of the Macroscopical Signal. Because his conviction that technocracy and globalising economic growth cause today's unsustainability, his response is to forcefully push back both. Yet I think the Signal demands adaptation through massive innovation; massive adaptive innovation so to speak. And innovation thrives best in free markets, stimulating creativity and generating the financial means to invest in research, development, demonstration and deployment – provided proper governmental preconditions restrict the 'innovation space' to sustainable trajectories. The Arcadian idea, that a global revolution from unsustainable once-through to sustainable closed-loop operation can be accomplished with existing or emerging technologies and a 1 kW energy budget per capita, is simply naïve, all the more closing material loops costs energy. In my view the development of a resilient sociometabolism is not *only* a socio-economic issue of wealth distribution, but *also* a grand techno-scientific challenge to sustain wealth creation.

THE NEXT ENERGY REVOLUTION

Symbian Man

Do you remember these words by Richard Fortey [29d]?

> "... gasp by gasp, oxygen was added and carbon dioxide commensurately reduced. It was life processes that shaped the atmosphere, that paved the way for other, more advanced organisms."

The blue-greens, rulers of the Phototrophic Regime, surprised themselves with an oxygen curse. Gasp by gasp they added oxygen until it began to affect the carrying capacity of the Earth for their own population and for almost all their co-inhabitants. Now look at humankind today. Gasp by gasp, carbon dioxide is emitted. It is anthropogenic processes that shape the atmosphere. *Homo sapiens carbonius*, ruler of the Carbocultural Regime, begins to affect the carrying capacity of the Earth for his own population and most of his co-inhabitants.

Of course, the comparison is metaphorical. Climate change is 'only' one of the multitudes of natural and socio-economical signals building up today. But my reference to the blue-greens is to illustrate two points. In the first place, these creatures demonstrated that even microorganisms can change the evolutionary course of 'system Earth' revolutionarily – if only uncountable individuals synchronise their act. Secondly, the power to survive the self-induced oxygen crisis came through adaptive innovation enabled by symbiotic partnerships, partnerships that still grow today's aerobic species in the 'Symbiotic Planet', including ours [51]. I wonder, perhaps symbiosis could well be a powerful response to today's macroscopical signals of unsustainability, too – yet not at the microbial level, of course.

The Emerging Helio-Energy Revolution

Today 'system Earth' can be seen as being composed of an ecosphere and an anthroposphere. The ecosphere comprises living and non-living subspheres, or the biosphere and the geosphere. Likewise, the anthroposphere consists of a noösphere and an artosphere. The anthroposphere is about 2.5 million years young. Still, it established ecological dominancy and, within it, the noösphere became leading. Thus, in the noösphere, or the mindsphere, or the sphere of thoughts created and shared by human beings, lies the key to sustainable development. The memes must do it – do not count on the genes, as the biologist Richard Dawkins explains [159]:

"There is something profoundly anti-Darwinian about the very idea of sustainability [as] short-term genetic benefit is all that matters in a Darwinian world ... Humans are no worse than the rest of the animal kingdom. We are no more selfish than any other animals, just rather more effective in our selfishness and therefore more devastating. All animals do what natural selection programmed their ancestors to do, which is to look after the short-term interest of themselves and their close family, cronies and allies ... The only solution to the problem is long-term foresight, and long-term foresight is something that Darwinian natural selection does not have ... [However, through] using the large brains ... it is possible to fashion new values that contradict Darwinian values ... hope lies in a uniquely human capacity for foresight."

Similar to 'natural selection' in biological evolution, 'cultural selection' in the evolution of human civilisation 'is not a process of chance and accident only, but also has deterministic elements' (as discussed in the Introduction). And just as 'natural selection' proceeds in two steps – firstly, genetic variation is produced and then selection operates on the emerged variants – 'cultural selection' follows the same route: firstly memetic and artefactual variation is produced and then selection operates on the emerged variants. In nature at the first step *"everything is a matter of chance"* while at the second *"chance plays a much smaller role"*, because *"the 'survival of the fittest' is to a large extent determined by genetically based characteristics"*. Likewise in human culture the 'survival of the fittest' is to a large extent determined by memetical and artefactual characteristics. Both 'natural selection' and 'cultural selection' can be seen as a process of elimination or 'differential survival'. Perhaps the most essential distinction is that perceptions of reality make a difference, and hence the human capacity for foresight. For the first time in the history of life a distinct species has an evolutionary choice to make – and not choosing is not an option. The species in question is *Homo sapiens* and the choice is to respond or not to respond to the macroscopical signals of unsustainability – and of course, if yes, how?

For sustainable development to become real, a new perception of reality – other than the Imperial or the Arcadian – must become dominant. Let me recall the rhythm of progress

discovery – invention – innovation – diffusion.

From this rhythm symphonies arise in the four parts

human knowing – human capacity – human acting – human living.

The compositions emerge from

new observation – new creation – new practice – new way of living.

At this juncture the 'observation revolution' marked by the Macroscopical Signal is inducing a scientific revolution. In a way the Signal triggers a reversal of the Copernican Revolution, which was sparked off by the inventions of the microscope and the telescope, in fact the first 'observation revolution'. This second Copernican Revolution, as Hans-Joachim Schellnhuber says, *"will enable us to look back on our planet to perceive one single, complex, dissipative, dynamic entity, far from equilibrium"* [68]. New 'macroscopical sciences' are emerging, such as Earth System Analysis, Industrial Ecology and Evolutionary Economics. All aim to get to grips with the elusive complexity and dynamics of eco- and sociosystems, as well as 'system Earth' as a whole. These complexity sciences emerge from a symbiosis between scientific reductionism and scientific holism,* or simplicity and complexity. Says E.O. Wilson, in his beautifully worded text [69c]:

> *"Complexity is what interests scientists in the end, not simplicity. Reductionism is the way to understand it. The love of complexity without reductionism makes art; the love of complexity with reductionism makes science."*

The new sciences seek *"consilience"* by uniting knowledge of natural and cultural disciplines, so leaving the science schism of the 19th century behind. Only through consilience has humankind a chance to understand 'system Earth' phenomena from the nano- up to the macroscale.

The Macroscopical Signal could trigger a 'memetic symbiosis' between Imperial Man and Arcadian Man, a symbiosis between not only reductionistic and holistic thinking, but also between anthropocentrism and ecocentrism, between *Homo economicus* and *Homo ecologicus*, and between

---

*I explicitly talk about 'scientific holism', which is in a glaring contrast with all sorts of New Age denotations of 'holism'.

The Emerging Helio-Energy Revolution

'Nature mastery' and 'back to Nature'. Just as with the 'genetic symbiosis' of about two billion years ago between photosynthesising and aerobically respiring microorganisms, the 'memetic symbiosis' between the Imperial and Arcadian worldviews could turn degradation into progress along new, more sustainable evolutionary paths. And 'memetic symbiosis' is more than two cultural species learning from each other; it is a synthesis of ideas evolving its own way. Where genetic symbioses struggle for survival in the natural world, memetic symbioses fight for existence in the cultural world, albeit both worlds co-evolve.

I have named the newly emerging memetic species procreating the symbionic worldview Symbian Man. Symbian Man's perspective on life is neither *anthropocentric*, nor *ecocentric*, but *'symbiocentric'* (see Figure 7.1*). As *Homo energeticus* he views the eco- and sociosystems making up 'system Earth' as co-evolving energy-dissipating structures. In his sustained effort to reach consilience, Symbian Man unites techno-scientific and socio-ethical virtues of both his memetic ancestors. He is both aware of his countless lifelines with nature and at the same time he strongly values the uniqueness of human culture. Symbian Man understands the inherent uncertainties of evolutionary processes, whether natural or cultural. Arcadian Man wants to go back to nature, but which nature? Perhaps the only certainty is that evolving backwards is not an option (a corollary of the Second Law, says *Homo energeticus*). Imperial Man wishes to master nature, but how? How do you control the unpredictable and inherently unknown?

The Helio-Energy Revolution

Symbian Man interprets sustainable development as nothing less than *orchestrating* the Helio-Energy Revolution after the *just happened to happen* Pyro-, Agro- and Carbo-Energy Revolutions. According to the Staircase model six revolutions need to operate in concert. In present-day parlance, this implies a revolution in human consciousness followed by the quartet of a science, a technology, an economic and a societal revolution, with an energy revolution at the core.

In essence, a heliocultural revolution contours a sustainable world with a human footprint fitting within the carrying capacity of the Earth and an equitable distribution over the global population. In helioculture the dominance of reductionism in knowledge creation is broken by macroscopical,

---

*See for the 'Arcadian perception' pictured Herman Daly's 'Beyond Growth: The economics of sustainable development' [157b].

Energy – Engine of Evolution

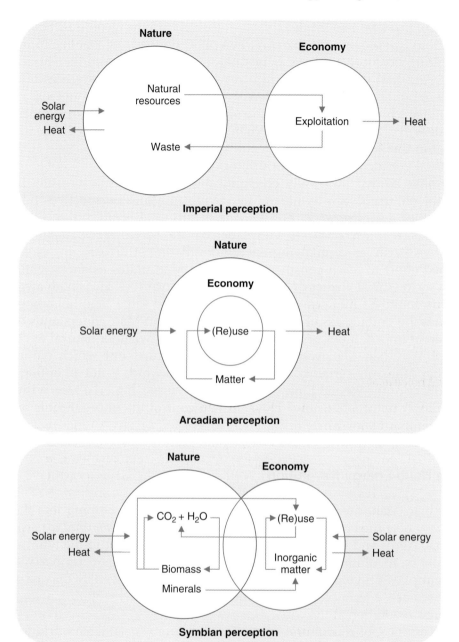

Figure 7.1 Perceptions of 'system Earth'

or metabolic, or complexity, or systems thinking; human capacity is aimed at making closed-loop technologies enabling the solar-driven recycling of matter; human acting is focussed on innovation through tripodal – people, planet, profit – value creation;* and societal structures, both governmental and non-governmental, are centred around partnerships of all kinds, from local to regional to global. Still, if helioculture develops into the next ecologically dominant energy regime, uncountable manifestations must be possible. That is why I do not try to characterise the next steps of the Staircase with key words like I did for the pyro-, agro- and carboculture – except for the domain of human knowing, for which I write *systems thinking* (see Figure 7.2).

Then, what can be said about the next energy revolution, except that it is beyond compare? Vaclac Smil writes [134h]:

> *"Beyond the fossil fuels the world can tap several enormous renewable flows: direct solar radiation and wind energy in the accessible layer of the troposphere are both several orders of magnitude larger than the current global total primary energy supply and they can be supplemented by hydroenergy and geothermal flows."*

According to Smil's judgement:

> *"The transition from societies energized overwhelmingly by fossil fuels to a global system based predominantly on conversions of renewable energies will take most of the twenty-first century. A very long road lies ahead of the renewables."*

Unlike *Homo ecologicus* and like *Homo economicus*, *Homo energeticus* sets himself the task of increasing energy flows through ground-breaking innovation, while like *Homo ecologicus* and unlike *Homo economicus*, *Homo energeticus* focusses on renewables and recycling. Of course, *Homo energeticus* aims at the highest energy utilisations possible, as harvesting solar energy is far from easy, but, more importantly, he is fully aware of the sheer size of the human enterprise in metabolic terms. The question is whether 'system Earth' can sustain a second giant energy-dissipating

---

*Ecological economists discuss concepts such as poor and perfect substitutability in relation to the notions of weak sustainability and strong sustainability, respectively [160]. Weak sustainability *"expresses the idea that the sum of man-made capital and natural capital should at least be kept constant"* assuming *"perfect substitutability between natural goods and man made goods"*. Strong sustainability assumes *"that natural capital and man-made capital are poor substitutes for each other"*. This way the discussion about sustainable development boils down to the question of substitutability between natural and created capital [151a]. However, thinking in terms of substitutability implies natural capital to be monetarily quantifiable. But natural functions emerge from ecosystems that behave inherently unpredictably and flourish largely beyond the horizon of human knowledge. In consistency, the many cost-benefit techniques each produce different outcomes [161a]. Therefore, I honestly doubt the usefulness of the substitutability concept, and, accordingly, the notions of weak and strong sustainability.

Energy – Engine of Evolution

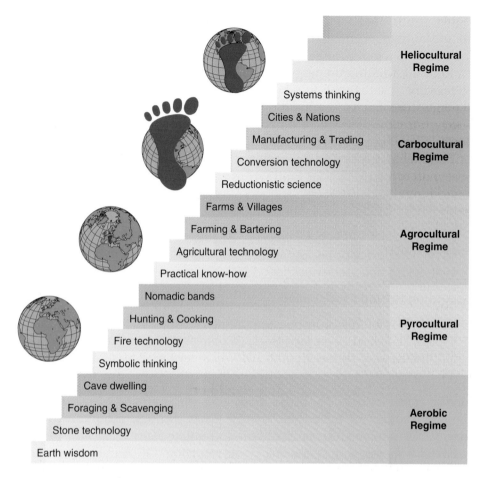

Figure 7.2   The emerging Heliocultural Regime

structure of a few dozens of terawatts in addition to the ecosphere processing a 100 TW flux. According to *Homo economicus* the answer is 'definitely yes', while *Homo ecologicus* says 'absolutely not' (he draws the line at about 8 TW as discussed before); but *Homo energeticus* thinks 'yes, provided that …' (some key characteristics of Imperial, Arcadian and Symbian Man are presented in Table 7.1).

During the Pyroic Era genus *Homo* evolved from early fire master to agricultural fire master to industrial fire master. He began as a fire master and he still is a fire master. Now his sole surviving species, *Homo sapiens*, has the mammoth task of transforming the fire mastership that made his genus human into new, true energy mastery. Such a shift will mark a new energy era whereby not only ecosystems, but also sociosystems harvest, store and utilise incident sunlight. If the Helio-Energy Revolution

Table 7.1 Three cultural, or memetic, subspecies of *Homo sapiens*

	Imperial Man	Arcadian Man	Symbian Man
Cultural species	*Homo economicus*	*Homo ecologicus*	*Homo energeticus*
Philosophy	Anthropocentric	Ecocentric	Symbiocentric
Future image	Nuclear Valley	Green Valley	Sun Valley
Energy regime	Nucleocultural	Ecocultural	Heliocultural
Energy source	Nuclear	'Conservation'*	Solar
Energy carriers	Electricity, hydrogen	Electricity, biofuels	Electricity, hydrogen, biofuels
Cultural approach	Technocratic	Socio-ethical	Socio-technological

happens, the Helian Period and the Helioic Era may be added to the Energy Time Scale. Socio-technological revolutions take time, although during the history of human civilisation the pace has accelerated: the Pyro-Energy Revolution took hundreds of thousands of years, the Agro-Energy Revolution tens of thousands of years and the Carbo-Energy Revolution hundreds of years. In view of the surging Carbian Explosion, Symbian Man is certainly striving for an even more rapidly evolving Helio-Energy Revolution, but not with the idea that sustainable development implies a continuously increasing 'revolution frequency'; on the contrary, the Heliocultural Regime should bring quietness to Sun Valley.

Sun Valley

Symbian Man sets course for Sun Valley where a Heliocultural Regime runs a closed-loop economy with integrated and cascaded flows of renewable energy and recyclable matter. The energy carriers are green electricity, solar hydrogen and green biofuels. (If biofuel manufacture consumes fossil or nuclear energy, the biofuel in question is not wholly renewable; therefore I add 'green' to biofuel.) The beauty of green biofuels is that Nature looks after carbon recycling through photosynthesis, with energy storage for free. And albeit that efficiencies are relatively low, residues of food and wood production nevertheless grow. In Sun Valley socio-metabolisms could differ regionally with ecological circumstances and heliocultures. They show optimised configurations of large-, small- and also medium-scale solutions (while Imperial Man prefers the large scale, and Arcadian Man the small). Generally, to improve the resilience of economies in terms

---

*Conservation is not an energy source. Arcadian Man opts for solar energy as the prime driver of green development, but using less energy is more characteristic of ecoculture.

of safety, security and above all adaptability, a Symbian approach favours distributed small-scale and decentralised medium-scale socio-metabolic sites and corresponding infrastructures. Needless to say, Symbian Man carefully maintains a competitive edge as well as an adaptive edge; hence he continuously improves his macroscopes.

Symbian Man hears the rhythm of progress – *discovery – invention – innovation – diffusion* – and views sustainable development as a grand socio-technological challenge (or actually a techno-scientific socio-ethical and corresponding socio-economical challenge, but that is quite a mouthful). His task is to utilise significantly more energy than carboculture today by increasing energy efficiencies and enhancing solar energy harvesting, but also by recycling matter with renewable energy; in effect, a double trend break with carbocultural practice. Building the capacity to supply one or more dozens of terawatts of renewable energy needs a dedicated and sustained innovation effort. A heliocultural closed-loop economy will be more complex than the current-day's carbocultural metabolism, but also, as with all energy revolutions, new uniformity could emerge – I think of Nature which has selected a universal fuel for cell metabolism, that is adenosine triphosphate (ATP).

For the next half century, humanity will try to reduce poverty and improve living standards, especially in the developing world, while combating climate change, biodiversity decline and environmental degradation. In the mean time the population will grow by perhaps three billion people, which will bring a concomitant increase in metabolic activity in terms of organic and inorganic material, water, food and energy flows. Rapid metabolic innovation is urgently needed, but at the same time difficult because of its inherent intertwining with all human activities; and once emerged, an energy culture resists change. Therefore, Symbian Man focuses on leapfrogging to heliocultural economies in developing countries, while reinventing carbocultural socio-metabolisms in the industrialised world.

Most likely, fossil energy will be used at a large scale to fuel socio-technological development for many years to come. Yet the innovation force required to develop economies sustainably with 'ancient sunlight' can only be underestimated. Many emission-related issues accompanying fossil fuel combustion, such as local health problems, regional environmental degradation and global warming, necessitate innovation efforts directed towards 'emissions control'. The options are efficiency improvement and clean metabolic routes, where 'clean' means near-zero emission. As well as 'conventional toxics' such as sulphur and nitrogen oxides and soot, the greenhouse gas carbon dioxide figures now on the list. To suppress its discharge into the environment, Symbian Man seeks solutions through, for example, geological and chemical sequestration

[162]. The latter conversion yields stable carbonates that can be 'recycled' in the form of construction materials, or otherwise.

Socio-metabolic complexes

To illustrate metabolic thinking as well as a contextual approach to innovation, I would like to round off with a Symbian idea. By 'contextual' I mean: fitting within the context of 'our common purpose', that is sustainable development, understood as travelling from Carbon Valley to Sun Valley, and aimed at reducing and simultaneously equalising humanity's ecological footprint.

In the coming decades the human population is expected to grow largely in urban(ising) areas of developing economies. So there the need for food, water, energy and construction materials will burst. However, leapfrogging to a solar-driven closed-loop metabolism in one go is simply impossible. A more sensible approach is to look for 'leapfrogging pathways'. Shaping such 'leapfrogging pathways' is a global challenge, all the more so when the sustainable solutions developed could be applied to reinvent today's carbocultural socio-metabolisms.

Metabolically speaking, the carbocultural way is characterised by large-scale remote processing and correspondingly extensive distribution networks, such as those used for electricity, drinking-water, sewage, food, transportation fuels and natural gas. One consequence is that energy and matter flows are being optimised in industrial isolation: oil refineries optimise theirs, and so do the pharmaceutical, automotive, steel and cement industries, to name but a few. The cement industry* thinks of entering energy and carbon dioxide markets with 'cement-industry-borne' services [163a,b], while the oil industry considers delivering 'oil-industry-borne' construction materials. Likewise, the agricultural industry aims to reduce the production of waste, while the energy industry explores routes for manufacturing biofuels from agricultural residues.

However, by optimising single supply chains a closed-loop economy will never be achieved, so a Symbian approach addresses a societal metabolism as a whole system. Only through integrating and cascading energy and matter flows could radical improvements in resource efficiencies be realised. An obvious starter is to address the energy inefficiency of carbocultural power supply. The dominating coal-fired power plants reject about two-thirds of the energy input. State-of-the-art

---

*The cement industry emits approximately five percent of global, man-made carbon dioxide emissions [163c].

Energy – Engine of Evolution

gas-fired power plants perform better, but still roughly half of the fossil energy ends up heating the open air. Integrating heat and power production clearly seems the way forward. But as heat cannot be distributed over long distances, combined heat and power (CHP) implies decentralised, or distributed generation, which means a radical shift in thinking and acting. Still, distributed power is not a systemic, or metabolic, approach to resource optimisation. That would require incorporating other societal lifelines, such as water, wood, food, construction materials and waste streams. Notably, large industries today do show a metabolic tactic, but obviously only within their industrial fences. A modern refinery, for example, is not one plant, but a complex of dozens of plants feeding each other with their respective outputs, including, for example, a generator converting waste heat into power and a biological treater detoxifying industrial water. The time is now ripe to apply industrial knowledge and know-how on integrating and cascading energy and matter flows to socio-metabolisms of, for instance, a city. This symbiosis between city and industrial development could lead to complexes of 'socio-metabolic sites' with, for example, in addition to combined heat and power, a sewage treater, a bio-residue-to-fuels manufacturing plant, a natural-gas-to-hydrogen converter or a carbon dioxide mineralisation plant (see Figure 7.3). The modules of such a socio-metabolic complex, what we call a CityPlex, can be standardised to a large extent. Flexibility resides in the modular composition being adaptable to its cultural environment.

A city metabolism cannot easily be cyclic in itself; that is because energy and matter flows, including food and water, usually come from outside.

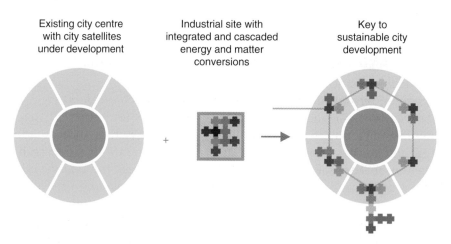

Figure 7.3   The Symbian CityPlex Concept

The Emerging Helio-Energy Revolution

Whereas improving resource efficiencies may possibly favour small- or medium-scale socio-metabolic sites inside cities, outside the large-scale might be better. For example, large-scale coal processing with large-scale carbon dioxide production could enable large-scale geological sequestration. And that would bring with it large-scale fossil hydrogen generation, which, in turn, may well be best applied in small- or medium-scale combined heat and power CityPlex units – a Symbian approach seeks optima. Besides carbon dioxide and fossil hydrogen production units, a CoalPlex could be equipped with a desulphuriser plus a downstream sulphur cement manufacturing unit and a synthetic transportation fuel plant. Of course mobility economies need to be included in the whole picture, as do information and communication systems. In rural areas, socio-metabolic complexes can be envisioned that feed on biomass and other renewable sources. Efficient 'metabolic networking' between villages in the form of RuralPlexes could open routes to both sustainable land development and food, wood and biofuel supply to urban areas, provided nutrient loops are closed. The latter requirement could necessitate the recovery of phosphorus and other mineral elements from municipal and industrial waste streams for reuse in fertiliser manufacturing.

Integrating and cascading energy and matter conversions is the way to a closed-loop economy.* Resulting socio-metabolic complexes such as CityPlexes, CoalPlexes and RuralPlexes could thus become relatively resilient stepping stones on 'leapfrogging pathways' to Sun Valley. Their creation demands innovation coalitions, or Symbian partnerships, between governments, non-governmental organisations, universities and companies in research, development, demonstration and deployment – on the four levels of the Staircase of Socio-Technological Development – and also between East, West, North and South.

AND THE FACE OF THE EARTH WILL CHANGE

I guess that a space traveller in 2050 looking at planet Earth will observe a messy mix of Carbon Valley icons, with niches of Nuclear Valley, Green Valley and Sun Valley. If these niches stay small or even invisible, then the traveller may see Carbon Valley in a state of decay, because the Earth could no longer carry the Carbian Explosion. But if the niches do grow, then the extraterrestrial spectator will see an evolutionary battle between

---

*Industrial Ecology is an emerging academic discipline aimed at achieving just this, although the initial approach is more academic than industrial and needs, in my view, industrial strengthening.

Imperial Man, Arcadian Man and Symbian Man. Maybe the next dominant energy regime will already be apparent. However, in this century without a doubt the face of the Earth will change, and rapidly it seems, for the human force is unmatched. Either humankind itself or 'system Earth' will tame it, and 'system Earth' knows no boundaries.

It is the surging Macroscopical Signal telling me, and many others, that the Earth cannot sustain our carbocultural way of life for too long. The Signal prepares, as it were, people's minds for a new perception of reality, and I sincerely hope that it will be a Symbian perception. To me, the Signal itself is a fantastic achievement for human ingenuity, and not a Signal of doom and gloom. It is a Signal of choice and chance that can be seized or not. Humanity has a choice to make, an evolutionary choice. It has a chance to consciously switch from industrial development to sustainable development, or from carboculture to helioculture.

Perhaps, as experienced fire masters, we stand on the threshold of becoming true energy masters. But let us accept that we are only apprentices. If we manage to control human culture through socio-technological innovation, then sustainable development could become reality. Energy flows drive the evolution of the living planet, but we, *Homo sapiens*, could purposely influence its shaping.

– Appendix –
Energy, Complexity and Evolution

THE NATURE OF ENERGY

All evolutionary change is driven by energy flows, including the creation of living structures. But *how* does a living structure come into being? To address this quest, we must first understand the nature of energy and the way it supports the organisation of physical, biological and societal structures.

Classical mechanistic questions about the workings of energy bring us into the domain of thermodynamics. In thermodynamics, energy is *"the capacity to do work"* [100a], or the capacity to cause change – *physical* change, I hear my physicist friends, who always protect the strict severity of reasoning, adding. But all change has a physical basis, from the origin of species to the words we speak, and from the falling of an apple to the thoughts we think; thus the addition of the term *physical* is correct, albeit redundant. Classical thermodynamics emerged from the need to improve the steam engine, or, more accurately, to convert heat into mechanical work, and evolved into *"the study of the transformations of energy in all its forms"* [100a]. But still, say Lynn Margulis and Dorion Sagan [7n], thermodynamicists have focussed largely on *"machines or chemicals in closed boxes"* avoiding *"wet and wiggly"* creatures; and obviously *"living beings are not steam engines or closed mechanical systems"*. However, they observe that:

> *"We are in the throes of a great breakthrough: the linkage of the exact science of thermodynamics with other sciences such as biology, meteorology, and climate change ... The new thermodynamics of life is likely to have a lasting influence on evolutionary thought."*

Margulis and Sagan seem to sense the rise of a new uniting of knowledge, of new consilience between energy science and evolutionary science [7n]:

> *"Evolution is a science of connection, and connection does not stop with ties of humans to apes, apes to other animals, or animals to microbes: Life*

and nonlife are also connected in fundamental ways. We have seen that the organization of life is material. But energy also organizes."

Recently, scientists studying complex systems, such as 'system Earth', have begun to take classical thermodynamics beyond its original limits to describe the relationship between energy flow and the emergence of complex structures. I will try to explain a number of nascent insights, in particular those used in constructing the Staircase of Energy Regimes (see Chapter 6).

Let me begin with the foundation of classical thermodynamics, the Laws of Thermodynamics. They have proven to be robust within their validity domain, the physical and chemical systems for which they have been formulated. There are four laws, from the Zeroth to the Third [100a], but for the purposes of this investigation into the world of evolutionary change, we can limit ourselves to the First and the Second. The First Law can be articulated in normal language as *"Energy is conserved."* But what is 'conservation'? We need the Second Law to explain this. The Second Law adds that for change to occur, energy must transform into another quality. 'Energy quality' in this context is the real capacity to do work; it is the available energy, or the free energy. Take the example of driving a car. Eventually, after a ride, the energy stored by the fuel in the tank is dissipated as heat into the environment through radiation and friction. Thus, high-quality energy stored in the form of a fossil fuel is converted into low-quality energy dissipated in the form of heat; the total energy is conserved, but transformed. The moral is that for physicochemical change to occur, 'energy quality' has to be paid.

In effect, the Second Law expresses a fundamental asymmetry in nature, articulated by Atkins as follows [100a]:

"Hot objects cool, but cool objects do not spontaneously become hot; a bouncing ball comes to rest, but a stationary ball does not spontaneously begin to bounce."

In both cases, energy dissipates as heat to the surroundings, which act as a sink. Conversely, the concentration of energy never* occurs spontaneously. We observe the same asymmetry for matter. Air, for example, automatically leaks out of a balloon, which never blows up of its own accord. Atkins typifies the asymmetry [100b]:

"The natural tendency of energy to disperse – that is, to spread through space, to spread the particles that are storing it, and to lose the coherence

---

*Formally, it should be 'rarely', and the larger the system, the more rare the event.

Appendix: Energy, Complexity and Evolution

> *with which the particles are storing it – establishes the direction of natural events."*

Terms such as dispersion, or loss of coherence, describe a process from more order to less order, or from less chaos to more chaos, or – with a sense of drama – from order to chaos. The concept of entropy measures the level of chaos in energy and matter. Thus high-quality energy corresponds to low entropy, and low-quality energy corresponds to high entropy [100b]:

> *"High-quality energy must be undispersed energy, energy that is highly localized (as in a lump of coal or a nucleus of an atom); it may also be energy that is stored in the coherent motion of atoms (as in the flow of water)."*

Low-quality energy manifests itself in particular as heat at near ambient temperatures. The Second Law tells us that spontaneous change goes with the fall of energy from a higher to a lower quality, or rather from a lower to a higher entropy. In natural processes energy is conserved, but entropy increases. The nature of energy, it appears, is to decay, to dissipate, to produce chaos, or entropy. Rudolf Clausius (1822 to 1888), one of the bright minds pioneering thermodynamics, coined the term entropy to denote energy's *"tendency to dissipation"* [67b]; the word is derived from the Greek 'trope' for transformation [53l], or change.

Having the tendency to dissipate is one thing, but being able to dissipate is another. Energy needs dissipative paths in the physical world to produce entropy. But these paths are not always there. A so-called closed system, or a system that does not exchange energy or matter with its environment can adopt an equilibrium state, that is a state without change – thus without entropy production; it is a dead state, so to speak. A corollary of the Second Law is that a closed system in equilibrium will not spontaneously leave that condition (or at least, formally, the chance that it will is extremely small); and if it is not in the equilibrium condition, it tends to reach it. For example, when we bring a hot iron bar into an insulated house at ambient temperature (the room and the bar form a closed system), at a certain moment the room and bar temperatures will be equal. Then the house/bar system is equilibrated; there is no further spontaneous change. But if the house is not perfectly insulated, the house/bar system will equilibrate with the environment, adopting the outside temperature. Insulation aims to limit energy dissipation.

I would like to focus, for a moment, on a reading lamp. Perhaps you are using one right now to illuminate this text? No doubt, the lampshade feels warm. It has adopted a temperature in between that of the light bulb and the room; the resulting temperature depends on the temperature of the bulb, the material composition of the lampshade, and the temperature in the room. The lampshade temperature is constant, but still, there is continuous change in the form of light and heat production. As a consequence, the lampshade is not in equilibrium, but in a steady state: that is, it radiates the same amount of heat into the room as it absorbs from the light bulb, thus dissipating a steady flow of energy. The lampshade is thermodynamically speaking an open system; energy enters the bulb through the electric cabling and dissipates into the room. The difference between *equilibrium*, a concept reserved for a distinct state of a thermodynamically closed system, and *steady state*, a concept used to describe a certain condition of a thermodynamically open system, is very important although often overlooked. For instance, the saying that the Earth is, or strives to be, in equilibrium, is formally incorrect. Obviously, planet Earth is thermodynamically an open system: the Sun continuously hits it with a flow of solar energy. The Earth is considerably cooler than the Sun, and substantially warmer than its outer space. Hence the Earth continually seeks out a steady state from which it radiates the same amount of energy into outer space as it absorbs from the Sun. And just like the lampshade in the simple example above, the Earth adopts a temperature in between that of the Sun and the outer space which also depends on its material composition. Note that the quality of the incoming solar energy differs significantly from the quality of the outgoing radiation. What comes in is short-wave, high-quality, low-entropy sunlight, and what goes out is long-wave, low-quality, high-entropy infrared radiation [33r].

Not all sunlight striking the Earth enters into the atmosphere. Clouds, chemicals, dust and the Earth's surface reflect about 34 percent directly back into space [39d]. The remaining 66 percent of incoming solar energy heats the atmosphere and the Earth's surface (42 percent), evaporates water (23 percent), generates wind (1 percent) and fuels photosynthesis (0.023 percent). Solar energy trapped by the atmosphere is predisposed to disappear into the cold outer space, yet heat-trapping gases such as water vapour and carbon dioxide offer resistance. These gases, the so-called greenhouse gases, actually isolate the Earth through 'radiation insulation'. Without this atmospheric thermal blanket composed of greenhouse gases, the Earth's surface would be roughly 33°C colder [39f] with an average value of minus 18°C – probably too cold to nourish life; this natural 'greenhouse effect' makes the difference between a living and a lifeless planet. In every-day usage, the phrase 'greenhouse effect', or 'global warming', stands for the recent, suspected anthropogenic increase in atmospheric temperature, while formally this phenomenon is referred to as the 'enhanced greenhouse effect' [39g].

A closer look into the workings of absorbed solar energy reveals something strange. The transformation of sunlight into heat and water vapour is perfectly consistent with the Second Law: heat is a lower-quality energy form than sunlight and the evaporation of water goes with loss of coherence; thus, in both cases, entropy production occurs. But wind and photosynthesis yield order; relative order to be precise. Wind means coherent motion of air particles; photosynthesis yields living structures in the form of plants and trees which are composed of carbohydrates formed out of 'chaotically' dispersed water and carbon dioxide molecules. Wind and photosynthetic life correspond with entropy consumption instead of production. Are these processes then violating the Second Law? No, says James Lunine [17g], focussing on life:

> *"Life very definitely providing us with a "living" example of the Second Law, and is churning out entropy like crazy!"*

Lunine states that the Second Law allows entropy consumption, which is the creation of structure or the emergence of order out of chaos. But creating order does not come for free. The price for it is energy. In thermodynamic terms, a living organism is an energy-dissipating structure. Structure indicates order, and significant order indicates relatively low entropy. We have seen that, according to the Second Law, no system will leave the relatively high-entropy equilibrium state freely, but with 'energetic' help from outside it can do just that. When an energy input drives a system away from thermodynamic equilibrium, structure can emerge – energy-dissipating structure, to be clear. The maintenance of a far-from-equilibrium condition requires a continuous energy input. Hence a dissipative structure collapses if the sustaining energy flow fades; a living organism that stops eating dies. Equilibrium means death. Life lives far from equilibrium. But far-from-equilibrium is not synonymous with live.

ENERGY-DRIVEN ORGANISATION

The creative power of an energy flow manifests itself not only in living structures, but also in physical systems, such as whirlwinds [67c] or whirlpools [14f]. Everyone will be familiar with the fascinating vortex structure that suddenly emerges in bath water when the plug is removed. However common this vortex in the bathtub may be, in terms of energy management it represents an out-of-the-ordinary natural phenomenon. The vortex adds dynamic structure to the stagnant water mass from which it arose. However, as vortex order emerges spontaneously, the formation of a vortex seems to violate the Second Law. But this is not so. On the contrary, by accelerating the dissipation of bath water into the sewer, the vortex increases the rate of entropy production in the 'bathtub/sewer system'. What we observe in the bathtub is nature's creative power. Nature is able to create local structure to facilitate energy dissipation, provided the system being structured is in a far-from-equilibrium state. Thinking along these lines, researchers are discovering more and more instances of the spontaneous emergence of structure in nature, also called 'emergent properties' of complex systems. Prigogine (who was awarded a Nobel Prize for Chemistry in recognition of his contributions to non-equilibrium thermodynamics) describes the *"spectacular example of chemical oscillations"* that became known soon after he and his team had predicted the existence of spatiotemporal organisations in far-from-thermodynamic equilibrium conditions, and named them *dissipative structures* [164a]:

> *"I remember our amazement when we saw the reacting solution become blue, and then red, and then blue again. Today, many other oscillatory reactions are known, but the Belousov-Zhabotinski reaction remains historic because it proved that matter far from equilibrium acquires new properties. Billions of molecules become simultaneously blue, and then red. This entails the appearance of long-range correlations in far-from-equilibrium conditions that are absent in a state of equilibrium ... we can say that matter at equilibrium is "blind", but far from equilibrium it begins to "see." We have observed that at near equilibrium, dissipation associated with entropy production is at a minimum. Far from equilibrium, it is just the opposite. New processes set in and increase the production of entropy."*

The so-called 'Bénard instability' is an example of a physical dissipative structure [100c]. It *"forms in a liquid when convection occurs between two horizontal surfaces, the lower surface being hot and the upper cool. When the temperature difference between the two plates is low, there is a chaotic distribution of the moving particles of the liquid."* However, when the temperature difference is great enough, structure emerges spontaneously in the form of coherently moving molecules forming hexagonally shaped convection cells. When the heating process stops, *"the flows disintegrate and return to the usual thermal motion"* [164b]. Order out of chaos thus emerged on the wings of an imposed energy gradient. Likewise, ecological dissipative structures could, for example, emerge in the form of periodic oscillations of prey and predator populations [100c].

The term 'self-organisation' is often used to characterise this phenomenon of emergent properties, and, unfortunately, this term has become popular. I say unfortunately, indeed, wholeheartedly agreeing with Eric Chaisson [13b]:

> *"The term* [self-organization] *and others like it (those with the prefix "self-") are deceptive in that such ordering is actually occurring not by itself, as though by magic, but only with the introduction of energy."*

Appendix: Energy, Complexity and Evolution

Likewise, Margulis and Sagan emphasise that *"complex systems are ... not "self"-organized"* [7o]. For organisation to emerge, a system must first be pushed far from equilibrium under the influence of an energy flow. Then, *in* the system, *structure* could emerge spontaneously, seemingly 'self-organised', to increase the rate of energy dissipation through the structure. Hence, without exception, 'self-organisation' is energy-driven, and 'energy-driven organisation' would be a more accurate description with which to typify emergent phenomena. The *shape* of an energy-driven organisation will always depend on the nature of the interactions between its composing parts.

With regard to energy-dissipating structures, it is important to discriminate between *driving* forces and *shaping* forces: a driving force results from the influence of an energy flow through the system, and a shaping force results from the interactions between agents in the system. In the case of the vortex in the bathtub, the driving force is gravity, whereas shaping forces result from the interactions between individual water molecules. If you were to fill the bath with salad oil, you would see the emergence of a differently shaped vortex. The sustainability of vortices like whirlwinds and whirlpools is usually poor. They collapse almost as suddenly as they emerge. But when the supply of high-quality energy is sustained, they can remain 'forever'. For instance, with the tap on and the plug removed the water flow sustains the vortex structure.

The Laws of Thermodynamics also apply to chemical processes, or processes that change the molecular structure of matter. Within a chemical system an energy-dissipating structure can emerge, provided it is pushed far from equilibrium through a supply of energy. For example, the product of a chemical reaction can *auto*matically start to act as a *catalyst* to accelerate its own production, a phenomenon that is called *autocatalysis* [17g]. Autocatalysis is a form of *energy-driven organisation* at the molecular level accelerating entropy production. In effect, an autocatalytic system is a kind of invisible 'chemical vortex'. Under persistent exogenic pressure of energy and/or matter flows, autocatalytic systems can increase their level of complexity over time by evolving to a next energy-dissipating structure of more complex networks of chemical transformations. Thus, an early autocatalytic system not only represents the creation of new *chemical information*, it has the inherent capacity to *create* chemical information through chemical evolution. For example, the biochemistry of life shows a remarkable structure in the form of a rigorously enforced asymmetry: nearly all amino acids that take part are left-handed, while the chemical formation reactions show equal preference for the two mirror-image geometries [165a]. Although the true origin of this asymmetry remains an unsolved problem, it could only have been realised in a complex framework of dissipative structures with appropriate chiral autocatalysis.

In 'Steps Towards Life: A Perspective on Evolution', Manfred Eigen argues that autocatalytic chemical evolution is the ultimate cause of *genetic information*, in essence a form of *chemical information* [32d]. He maintains that *"cellular life appeared on our planet, after ... chemical self-organization"* [32e]. In 'At Home in the Universe; the Search for Laws of Self-Organization and Complexity', biologist and pioneer in the emerging science of complexity, Stuart Kauffman, explores the notion of energy-dissipating structures to investigate spontaneous order in the living world. All free-living systems, he says, are dissipative structures, or *"complex metabolic whirlpools"*, within which a *"flux of matter and energy through the system is a driving force generating order"* [14f]. According to Kauffman, efforts *"to establish general laws predicting the behaviour of all nonequilibrium systems ... have not yet met with success. Some believe that they may never be discovered."* But he takes up the challenge by simulating the evolution of so-called 'complex adaptive systems', his name for free-living energy-dissipating structures, on the computer. Although Kauffman begins ambitiously with the quest for the origin of life, he tells his reader beforehand that [14g]:

"Anyone who tells you that he or she knows how life started on the sere earth ... is a fool or a knave. Nobody knows. Indeed, we may never recover the actual historical sequence of molecular events that led to the first self-reproducing, evolving molecular systems ... But if the historical pathways should forever remain hidden, we can still develop bodies of theory and experiment to show how life might realistically have crystallized, rooted, then covered our globe."

For Kauffman 'life' and 'metabolism' are inextricably linked to each other. A search for the origin of life is a search for the origin of metabolism, the creative fluxes of energy and matter. He argues that [14h]:

"... there are compelling reasons to believe that whenever a collection of chemicals contains enough kinds of molecules, a metabolism will crystallize from the broth. If this argument is correct, metabolic networks ... can spring full-grown from a primordial soup. Order for free, I call it."

But there is no such thing as a free lunch. Order never comes for free, but costs energy. Kauffman views the first metabolisms as autocatalytic molecular networks, and thus emergent properties of non-equilibrium chemical systems [14h,i]. For him the secret lies in autocatalysis:

"The cell is a whole, mysterious in its origins perhaps, but not mystical. Except for "food molecules", every molecular species of which a cell is

Appendix: Energy, Complexity and Evolution

constructed is created by catalysis of reactions, and the catalysis is itself carried out by catalysts created by the cell."

For Kauffman *"there is reason to believe that autocatalytic sets can evolve without a genome"* (a collective term for the inheritable, or genetic material of an organism) [14j]. When compartmentalised, the autocatalytic sets yielded so-called protocells, which could reproduce with heritable variations without genome, the autocatalytic system itself serving as a source of *chemical information*. From a world of protocells, an ecosystem inevitably would emerge [14j]:

"In short, a molecule created in one protocell can be transported to other protocells. That molecule may promote or poison reactions in the second protocell. Not only does metabolic life begin with whole and complex; but all the panoply of mutualism and competition that we think of as an ecosystem springs forth from the very beginning. The story of such ecosystems at all scales is the story not merely of evolution, but of coevolution."

It is thinkable that Kauffman's protocell evolved into the self-reproducing biological cell, characterised by a chemical code to record its own blueprint [32f]; blueprint here refers to the genome. The blueprint is passed on to offspring during cell reproduction. A gene is composed of DNA molecules, the structure of which can change as a result of the interaction with mutagenic chemical impurities, or high-energy radiation, but also by errors in the biochemical copying process [17e]; errors are rare, but do occur. When alterations become hereditary, mutations are born. Mutations add information to the genetic information pool and in that way make life more complex.

In effect, a living cell is an energy-dissipating structure: food molecules in, waste molecules and heat out – a sort of 'biological vortex'. The cell's metabolism serves to maintain its far-from-equilibrium state. Apparently, there is such a thing as a relatively stable far-from-equilibrium state. Kauffman talks about an 'attractor state'. The attractor state is a 'dynamical steady state' that shows the tendency to counteract perturbations enforced from outside. But when counteracting no longer works – that is, when destabilising positive feedback loops persistently overtrump the stabilising negative ones – then the energy-dissipating structure could gradually, or more or less abruptly, jump from one to another attractor state. In essence, the evolutionary development of an energy-dissipating structure follows a trajectory of attractor states.

Nature prescribes the direction of this evolutionary development but not the entropy-producing paths. Dilip Kondepudi and Ilya Prigogine assert that energy-dissipating structures evolve unpredictably. They state that no general extremum principles exist for such systems that control them [165b]. However, they do note that an attractor state *"may progressively drive ... to higher states of entropy production and higher states of order"*, thus conditions with increased complexity [165c]. In other words, the next step in the evolution of an energy-dissipating structure may result in more dissipation and thus in more local structure; 'may be', because depending on the circumstances an energy-dissipating structure can also fall to a simpler state, or collapse.

So far we have discussed single 'energy vortices' of physical, chemical and biological nature: a whirlpool in the bathtub, an autocatalytic chemical system and the biological cell. Hermann von Helmholtz showed mathematically in 1858 that the vortices of an ideal fluid (a model fluid that lacks viscosity) [166]:

> *"Could collide elastically with one another, intertwine to form complex knotlike structures, and undergo tensions and compressions, all without losing their identities."*

This ability to build 'multi-vortex structures' also holds for collections of 'energy vortices', or energy-dissipating structures. Kauffman suggests how autocatalytic protocells could build 'proto-ecosystems', and nature demonstrates a baffling range of composite 'biological vortices'. An individual organism can be seen as a superstructure of 'cellular energy vortices', and also as an energy-dissipating structure; or an 'organismic energy vortex'. An ecosystem can be seen as a superstructure of individual organisms, but also as an energy-dissipating structure; or an 'eco-energy vortex'. The whole ecosphere can be seen a superstructure of ecosystems, and as an energy-dissipating structure in itself; or a 'global energy vortex'.

Eric Schneider and James Kay reason that if the ecosphere operates as an energy-dissipating superstructure, the science of energy, that is thermodynamics, must be applicable. However, ecologists are confronted with the fact that the Laws of Thermodynamics have been formulated by physicists and chemists for physical and chemical systems, or systems with only physicochemical interactions between the composing parts (like gravitational interaction, electromagnetic interaction, molecular interaction, etc). But how do you apply the Laws of Thermodynamics to ecosystems, for example, and describe prey/predator interactions within a food chain, in effect a dissipative pathway for the degradation of photosynthetically stored solar energy? Kay reasons as follows [167]. A thermodynamically open system can be pushed away from equilibrium by the input of free energy. But, because leaving the equilibrium state is unnatural – a corollary of the Second Law – the system offers resistance.

Appendix: Energy, Complexity and Evolution

There is only one solution, and that is to dissipate the energy as effectively as possible which requires the creation of energy-dissipating paths. According to Kay, the more free energy is pumped into the system, the more organisation or structure can emerge to dissipate the free energy. He notes that systems tend to get better and better in utilising resources for the enhancement of energy dissipation, and thus the complexification of their structure, but only up to a certain limit. Beyond a critical distance from equilibrium, the system gets overloaded, loses its organisational capacity, and becomes chaotic.

Kay and Schneider propose a restated, more generalised Second Law, so that ecosystems fall within the domain of validity [168]:

> *"As systems are moved away from equilibrium, they will utilize all avenues available to counter the applied gradients* [such as temperature difference, pressure difference, etc]. *As the applied gradients increase, so does the system's ability to oppose further movement from equilibrium."*

In other words, forced by the energy gradient imposed by the Sun, ecosystems tend to increase their level of energy utilisation, and thus organisation [169]. To verify this hypothesis, field ecologists initiated campaigns to measure surface temperatures of different ecosystems. The data show an unmistakable trend: the more developed the ecosystem, the colder the surface temperature, and the lower the energy quality of the reradiated energy. Tropical forests show superior ability relative to grassland and deserts *"to cool themselves"* [70]. In other words, the more developed the ecosystem, the higher the level of energy utilisation, or the more effective the reduction of the solar energy gradient, and the higher the entropy production [70,167]. Also, the diversity in both species and trophic levels is vastly greater at the equator, where more than 80 percent of incoming solar radiation hits the Earth [170]. The 'unmistakable trend' observed by field ecologists looks like the 'invisible hand' Kauffman sees in his computer simulations. *"Over evolutionary time"*, he reasons [14k]:

> *"Ecosystems may self-tune to a transition regime between order and chaos, maximizing fitness, minimizing the average extinction rate, yielding small and large avalanches of extinctions that ripple or crash through the ecosystem."*

Chaisson notes that, universally, the more complex and intricate an energy-dissipating structure is, the more energy intake (per unit mass) it needs to sustain [13c]. He reformulates the Second Law such that it can be used to express that structure emerging in a distinct system goes with entropy production outside the system equal to, or larger than, entropy consumption inside the system. What counts is that overall, in the

Universe, entropy is produced, or rather energy is dissipated. Chaisson states:

> *"Although no formal proof of this inequality exists (or is perhaps even possible), the absence of any experimental exceptions provides confidence in its wide scale validity."*

The problem is that large energy-dissipating structures are simply too complex to describe rigorously, with all interactions included. Think of the Sun/Earth system, with planetary interactions, decomposing bacteria, trees, tree frogs, and human beings. System scientists have found an elegant way to depict the behaviour of *whole* systems *without the need to include all interactions between all separate parts*. Their approach is to look at feedback loops, which could be either positive or negative. A positive feedback loop is a circular chain of effects increasing change; a negative feedback loop is a circular chain of effects that opposes change [128b]. An example of a positive feedback is the exponential growth of a population of species when there happens to be a surplus of food, space, and other resources. Homeostasis represents a typical example of negative feedback in biological systems. Feedback loops are *emergent properties* of energy-dissipating structures. They are properties of the *whole* system, not of the composing *parts*.

We have reviewed 'energy vortices' of a physical, chemical, biological and ecological nature. There is one family to add: the family of societal (or cultural, or human, or anthropogenic) energy-dissipating structures, or sociosystems.* Ecologists see sociosystems as ecosystems and treat them in the same way, but in my view that is not correct. Indeed, both are energy-dissipating structures. But ecosystems produce energy flows through biological food chains, whereas sociosystems dissipate energy through cultural energy chains. Ecosystems and sociosystems exhibit fundamentally different metabolisms: in an ecosystem the 'energy conversion sites' are biological species, whereas in a sociosystem the 'energy conversion sites' are technological artefacts.

As discussed above, in my view, it makes sense to discriminate between *driving forces* and *shaping forces*. Whereas a Darwinian approach focuses primarily on *shaping forces*, an energetics approach describes evolution from the perspective of *driving forces* in the first place. An evolutionary *driving force*, or an *energy flow*, always stems from an *energy gradient*, with the gradient simply and generally defined 'as a difference across a distance' [7n]. An *energy gradient* has the inherent capacity to generate an

---

*Biologists describe that animals too form social structures. I fully appreciate that notion, but in this work a 'sociosystem' is a human sociosystem, or rather an anthropogenic societal energy-dissipating structure.

Appendix: Energy, Complexity and Evolution

energy flow between an *energy source* and an *energy sink*, such as a heat flow between a hot and a cold body. If an energy flow pushes a certain system far from thermodynamic equilibrium, an energy-dissipating structure can emerge and evolve along a trajectory of dynamic steady states, or attractor states.

During 'operation' an energy-dissipating structure can produce a new energy source or a new energy sink. Also, when it collapses, as happens once the structure's life-giving energy flow comes to a stop, or destabilising feedback loops can no longer be counteracted, it might leave an energy source or energy sink to the environment. Most intriguingly, ecosystems and sociosystems appear to be able to 'discern' an energy gradient other than the one(s) producing the energy flow(s) that keep(s) them alive. The energy systems appear to possess the capacity to develop, when the evolutionary conditions are right, innovative source-to-sink conversion paths allowing the utilisation of the newly 'sensed' energy gradient. A resulting, novel and viable energy-dissipating structure is actually an 'energetical' mutant. While real conversion from source to sink always proceeds via physical, chemical or biological paths, eco- and sociosystems manage energy flows between individual 'conversion sites' in food chains or energy chains, respectively. Energy-dissipating structures show an evolutionary lineage: biological energy-dissipating structures descended from chemical ones (as proposed by Stuart Kauffman) and evolved into ecosystems from which societal energy economies emerged (as discussed in Chapter 6). Energy-dissipating structures fight an evolutionary battle. And the dominant energy systems rule the energy regime.

The whole of all physical, chemical, biological, ecological, and societal energy-dissipating structures on earth form 'system Earth'. For James Lovelock, 'system Earth' evolves like a living organism, which he calls Gaia [171a]. In terms of energetics, Gaia must be seen as a global energy-dissipating superstructure catching energy flows from the Sun, and beaming it back into outer space after the Earthly work is done [7n]:

> *"Life does not exist in a vacuum but dwells in the very real difference between 5800 Kelvin incoming solar radiation and 2.7 Kelvin temperatures of outer space. It is this gradient upon which life's complexity feeds."*

Imagine that your brain does not translate visual signals into tangible shapes, but translates them into intangible energy flows instead. Then you would see a dazzling turmoil of 'energy vortices', from minuscule to gigantic. Smaller energy vortices, such as ants or human beings, interact in many ways to form larger structures, such as anthills or cities. And if you were looking at the Earth from a space shuttle through a powerful macroscope, you would see a global web of interacting *"metabolic whirlpools."* 'System Earth', seen as an energy-dissipating superstructure, produces feedback loops linking natural life with human life, or the ecosphere with the anthroposphere, and all life with its environment. Both change-inducing positive feedback loops and counteracting negative feedback loops happen to emerge, exactly as with other living, energy-dissipating structures.

The role of energy is to *drive* evolutionary processes, whether natural or cultural by character. All change on Earth, all change both around us and within us, is driven by surging, sustaining and fading *flows of energy*. A surging *flow of energy* could give birth to new energy-dissipating structure; a sustaining *flow of energy* maintains it; and when it fades, the *flow of energy*, the very structure becomes extinct. I would like to end this Appendix with a short history of the idea of *energy flow*, written by Eric Chaisson [13d]:

> *"Three-quarters of a century ago, biometrician Alfred Lotka (1922) identified energy's vital role, with 'evolution … proceeding in such direction as to make the total energy flux through the system a maximum compatible with the constraints' (see also Odum 1983). Today we would suggest that these systems optimize their flow. In turn, Lotka acknowledged his thermodynamic forebears: 'It has been pointed out by Boltzmann [1844 to 1906] that the fundamental object of contention in the life-struggle, in the evolution of the organic world, is available energy. In accord with this observation is the principle that, in the struggle for existence, the advantage must go to those organisms whose energy-capturing devices are most efficient in directing available energy into channels favourable to the preservation of the species."*

Now, we know the superior efficiency of complex ecosystems at reducing energy gradients, because *"as measured by low-flying airplanes and by satellites, ecosystems are cooler when they are more mature and biodiverse"* [7o]. But the same low-flying airplanes and satellites observe another pattern for sociosystems, where cities with large economic diversity dissipate considerable amounts of rejected energy. These fiercely dissipating sociosystems disturb ecosystems, affecting their biodiversity, and thus their energy utilisation. Life is connected with nonlife and biology with human culture. Hence we must fuse insights from the natural sciences and the cultural sciences to better know 'system Earth' at this juncture. As it is said by Edward O. Wilson, *"the testing of consilience is the greatest of all intellectual challenges"* [69e]. Perhaps, today, the energetics approach to evolution, understood as co-evolving complex energy-dissipating structures of both biogenic and anthropogenic nature, could help in strengthening consilience.

References

1. "history" from *Encyclopædia Brittannica*, 2004
2. J. Goudsblom, 2001, *Stof waar Honger uit Ontstond: Over Evolutie en Sociale Processen*, Meulenhoff (34-5)
3. S.J. Gould in Introduction to [28] (a:x, b:vii)
4. E. Mayr, 2004, *80 Years of Watching the Evolutionary Scenery*, Science, 305 (46-47)
5. "evolution" from *Britannica Concise Encyclopedia*, 2004
6. E. Mayr, 2001, *What Evolution Is?*, Basic Books (a:284-92, b:118-20)
7. L. Margulis and D. Sagan, 2002, *Acquiring Genomes: A Theory of the Origins of Species*, Basic Books (a:9, b:6, c:208, d:96, e:139, f:97, g:ix, h:145, i:7, j:88, k:98-9, l:156-7, m:84, n:43-6, o:47-9, p:207-17)
8. R. Aunger, 2000, *Darwinizing Culture: The Status of Memetics as a Science*, Oxford University Press
9. R. Dunbar, C. Knight and C. Power, 1999, *The Evolution of Culture*, Rutgers University Press
10. L.A. Dugatkin, 2000, *The Imitation Factor: Evolution Beyond the Gene*, The Free Press (a:21-23; b:2-4; c:137, d:84, e:108, f:89, g:10-3, h:118)
11. K.N. Laland, J. Odling-Smee and M.W. Feldman, 2004, *Causing a commotion*, Nature, 429 (609)
12. I. Prigogine, 1977, *Time, Structure and Fluctutations*, Nobel Lecture
13. E. Chaisson, 2001, *Cosmic Evolution: The Rise of Complexity in Nature*, Harvard University Press (a:12f, b:61, c:52-6, d:134)
14. S. Kauffman, 1995, *At Home in the Universe: The Search for Laws of Self-Organization and Complexity*, Oxford University Press (a:viii, b:71-3, c:243, d:149, e:98, f:20-1, g:31, h:45-7, i:50, j:73, k:235)
15. J. Goudsblom, 1992, *Fire and Civilization*, The Penguin Press (a:1, b:12-6, c:38, d:18-20, e:35-9, f:40-1, g:27-30, h:195-6, i:54-6, j:185, k:152-4, l:167, m:197, n:194, o:214-5, p:83-5)
16. A. Crosby, 1997, *The Measure of Reality: Quantification and Western Society, 1250–1600*, Cambridge University Press (a:12-9, b:71-3, c:99-103, d:78, e:cover)
17. J. Lunine, 1999, *Earth: Evolution of a Habitable World*, Cambridge University Press (a:35, b:106, c:162-3, d:134, e:138-9, f:219-221, g:154-6)
18. G. Hogan, 1998, *The Little Book of the Big Bang: A Cosmic Primer*, Copernicus, an imprint of Springer-Verlag New York, Inc. (a:127, b:viii)
19. V. Smil, 2002, *The Earth's Biosphere: Evolution, Dynamics, and Change*, The MIT Press (a:97-8, b:18)
20. M. Rees in Foreword to [18] (viii)
21. C. Wills and J. Bada, 2000, *The Spark of Life*, Perseus Publishing (a:71-2, b:76-7, c:40)
22. E.G. Nisbet and N.H. Sleep, 2003, *The physical setting for early life*, in [47] (a:5-9, b:16-7, c:17-20)
23. E.G. Nisbet and N.H. Sleep, 2001, *The habitat and nature of early life*, Nature, 409 (1083-91)
24. C. de Duve, 1995, *Vital Dust*, Basic Books (a:7, b:20, c:21, d:65, e:132, f:165, g:139)

25. J. Raven and K. Skene, 2003, *Chemistry of the early oceans: the environment of early life*, in [47] (a:63, b:57, c:59-60)
26. G. Horneck, 2003, *Could life travel across interplanetary space? Panspermia revisited*, in [47] (110)
27. J. Stanley and G. Orians, 2000/2003, *Evolution and the Biosphere*, in *Earth System Science: From Biochemical Cycles to Global Change*, ed. by M. Jacobson, R. Charlson, H. Rodhe and G. Orians, Academic Press (a:30, b:35)
28. C. Zimmer, 2001, *Evolution: The Triumph of an Idea*, HarperCollins Publishers, Inc. (a:109, b:65-6, c:109, d:102, e:66-9, f:190-1, g:144-7, h:71, i:159, j:161, k:164-5, l:285-6, m:267, n:290, o:260, p:288-90, q:299, r:295, s:305, t:302-3, u:182, v:186, w:150)
29. R. Fortey, 1999, *Life: A Natural History of the First Four Billion Years of Life on Earth*, Alfred J. Knopf (a:40, b:44, c:41, d:50-1, e:66, f:61-3)
30. W. F. Doolittle, 1999, *Phylogenetic Classification and the Universal Tree*, Science, 284 (2124-8)
31. S. Lamb and D. Sington, 1998, *Earth Story: The Shaping of our World*, BBC Books (a:177-8, b:181-3, c:193, d:190, e:197, f:201-3)
32. M. Eigen, 1987/1996, *Steps Towards Life: A Perspective on Evolution*, Oxford University Press (a:36, b:17-18, c:12, d:27, e:48, f:9)
33. V. Smil, 1999, *Energies: An Illustrated Guide to the Biosphere and Civilization*, The MIT Press (a:38, b:42, c:44, d:80, e:105-7, f:108-10, g:126-7, h:121-3, i:130, j:143-5, k:176-9, l:133, m:136, n:139, o:194, p:149-52, q:xv, r:11, s:156-7)
34. G. Michal, 1999, *Biochemical Pathways: An Atlas of Biochemistry and Molecular Biology*, John Wiley & Sons, Inc. and Spektrum Akademischer Verlag co-publication (a:99, b:13, c:200, d:182-5; e:44, f:16)
35. C. de Duve, 1984/1987, *De Levende Cel: Rondreis in een Microscopische Wereld*, Natuur & Techniek, 1987; original publication *A Guided Tour of the Living Cell*, first published by W.H. Freeman and Company, 1984; then published by Scientific American Books, Inc. (a:110, b:167, c:148, d:154-6)
36. E. Chaisson, 2001, *Cosmic Evolution: The Rise of Complexity in Nature*, Harvard University Press (a:168, b:11-13, c:55, d:133, e:193, f:140-1)
37. T. Lenton, 2003, *The coupled evolution of life and atmospheric oxygen*, in [47] (a:39, b:35, c:40-1, d:38, e:46, f:42-3, g:44-5)
38. D.E. Canfield, 1999, *A breath of fresh air*, Nature, 400 (503-5)
39. G.T. Miller, Jr., 2000, *Living in the Environment: Principles, Connections, and Solutions*, Brooks/Cole Publishing Company (a:G1-14, b:24-7, c:722, d:86, e:123, f:162, g:501)
40. V. Smil, 1997, *Cycles of Life: Civilization and the Biosphere*, Scientific American Library (41-3)
41. D. Des Marais, 2000, *When Did Photosynthesis Emerge on Earth*, Science, 289 (1703-5)
42. J.D. Macdouglass, 1996, *A Short History of Planet Earth*, John Wiley & Sons, Inc. (a:51, b:42, c:104, d:125, e:226)
43. A. Knoll, 2003, *Life on a Young Planet: The First Three Billion Years of Evolution on Earth*, Princeton University Press (a:106, b:98, c:19-20, d:100-1, e:103, f:24-5, g:139, h:23, i:132-3, j:123-4, k:94, l:121, m:153, n:159-60, o:126, p:157, q:151, r:105-7)
44. D.J. Catling, K.J. Zahnle and C.P. McKay, 2001, *Biogenic Methane, Hydrogen Escape, and the Irreversible Oxidation of Early Earth*, Science, 293 (839-843)
45. T.E. Graedel and P.J. Crutzen, 1995, *Atmosphere, Climate and Change*, Scientific American Library (a:62-3)
46. "cell" from *Britannica Concise Encyclopaedia*, 2004

References

47. L. Rothschild and A. Lister, 2003, *Evolution on Planet Earth: The Impact of the Physical Environment*, Academic Press, an imprint of Elsevier (a:426 under *"prokaryote"*, b:419-28)
48. D. Jablonski, 2003, *The interplay of physical and biotic factors in macroevolution*, in [47] (241-2)
49. S. Pyne, 2001, *Fire: A Brief History*, University of Washington Press (a: xvi, b:25, c:3, d:122-3, e:68)
50. "Precambrian life" from *Encyclopædia Britannica*, 2004
51. L. Margulis, 1998, *Symbiotic Planet: A New Look at Evolution*, Basic Books (a:38, b:6)
52. B. Sørensen, 2000, *Renewable Energy*, 2nd edition, Academic Press (a:132)
53. V. Smil, 1991, *General Energetics*, John Wiley & Sons (a:69, b:111, c:238, d:170, e:156-60, f:181, g:150-1, h:175, i:224, j:310, k:229, l:5)
54. R.G. Klein with B. Edgar, 2002, *The Dawn of Human Culture: A Bold New Theory on what Sparked the "Big Bang" of Human Consciousness*, John Wiley & Sons, Inc. (a:29-30, b:7, c:63-4, d:36, e:76, f:143-5, g:97-8, h:23, i:70-4, j:110, k:122, l:131, m:146, n:157, o:235-6, p:262, q:194, r:153, s:233, t:264, u:240-5)
55. R. Lewin, 1993/1998, *The Origin of Modern Humans*, Scientific American Library (a:21-2, b:25-8, c:15, d:31-3, e:165-6)
56. T.W. Deacon, 1997, *The Symbolic Species: The Co-evolution of Language and the Brain*, W.W. Norton & Company (a:12, b:21-3, c:40, d:43, e:379, f:112-5, g:382-4, h:392-3, i:396-7, j:410, k:452, l:336, m:458-9)
57. A. Ronen, 1998, *Domestic fire as evidence for language*, in *Neandertals and Modern Humans in Western Asia*, ed. by Akazawa et al., Plenum Press
58. S. Pyne, 1995/1997, *World Fire: The Culture of Fire on Earth*, University of Washington Press (a:x, b:4, c:13, d:302-3)
59. S. Pyne, 1997, *Vestal Fire: An Environmental History, Told through Fire, of Europe and Europe's Encounter with the World*, University of Washington Press (a:16, b:25-6, c:59, d:50, e:360)
60. J. Goudsblom, 2002, *The Expanding Anthroposphere*, in *Mappae Mundi: Humans and their Habitats in a Long-Term Socio-Ecological Perspective*, ed. by B. de Vries and J. Goudsblom, Amsterdam University Press (a:28, b:36-9)
61. M. Fischer-Kowalski and W. Hüttler, 1998, *Society's Metabolism: Review of the Intellectual History of Material Flow Analysis Part II: 1970-1998*, Journal of Industrial Ecology, 2(4)
62. B. de Vries and R. Marchant, 2002, *Environment and the Great Transition: Agrarianization*, in *Mappae Mundi* [60] (a:97, b:71, c:74-5, d:77, e:102-3, f:99, g:109-10, h:80-1)
63. Johan Goudsblom and B. de Vries, 2002, *Towards a Historical View of Humanity and the Biosphere*, in *Mappae Mundi* [60] (a:34)
64. R. Marchant and Bert de Vries, 2002, *The Holocene: Global Change and Local Response*, in *Mappae Mundi* [60] (53-5)
65. C. Holden, 1995, *Very Old Tools*, Science, 295 (795)
66. C. Ponting, 1991, *A Green History of the World: The Environment and the Collapse of Great Civilizations*, Penguin Books (a:271, b:88-91, c:151, d:279-81, e:115, f:267 70)
67. M.J. Sparnaay, 2002, *Van Spierkracht tot Warmtedood: Een Geschiedenis van de Energie*, Voltaire (a:201, b:24, c:194)
68. H.-J. Schellnhuber, 1999, *'Earth system' analysis: and the second Copernican revolution*, Nature, 402 (C19-C23)
69. E.O. Wilson, 1998/1999, *Consilience: The Unity of Knowledge*, Vintage Books (a:31, b:49-51, c:54-9, d:164, e:12)

70. "history of technology" from *Encyclopædia Brittanica*, 2004
71. B. de Vries, 2002, *Increasing Social Complexity*, in *Mappae Mundi* [60] (a:181)
72. J.A. Tainter, 1988/2000, *The Collapse of Complex Societies*, Cambridge University Press (a:26-30, b:33, c:5, d:7-12, e:38, f:23)
73. R.P. Sieferle, 1982/2001, *The Subterranean Forest: Energy Systems and the Industrial Revolution*, The White Horse Press (a:125-30, b:110-1, c:132-5, d:98, e:41, f:78-9)
74. "Darby, Abraham" from *Brittannica Concise Encyclopædia*, 2004
75. D. Landes, 1998/1999, *The Wealth and Poverty of Nations: Why some are so Rich and some are so Poor*, W.W. Norton & Company (a:40)
76. "Industrial Revolution" from *Brittannica Concise Encyclopædia*, 2004
77. D. Yergin, 1991, *The Prize: The Epic Quest for Oil, Money, and Power*, Simon & Schuster (a:14)
78. J.S. Robbins, 1992, *How Capitalism Saved the Whales*, The Freeman, August 1992, by Foundation for Economic Education, Inc.
79. "Drake, Edwin Laurentine" from *Encyclopædia Britannica*, 2004
80. P. Freiberger and M. Swaine, 2004, "computers" from *Encyclopædia Britannica*, 2004
81. L.R. Brown, C. Flavin and H. French, 2001, *State of the World 2001*, The Worldwatch Institie, W.W. Norton & Company (a:6)
82. J. Alper, 2002, *The Battery: Not Yet a Terminal Case*, Science, 296 (1224-6)
83. J.-M. Tarascon and M. Armand, 2001, *Issues and challenges facing rechargeable lithium batteries*, Nature, 414 (359-67)
84. International Energy Agency (IEA), 2003, *Key World Energy Statistics 2003* (a:24, b:7)
85. "nuclear fission" from *Encyclopædia Britannica*, 2004
86. "uranium" from *Encyclopædia Britannica*, 2004
87. J. Perlin, 1999, *From Space to Earth: The Story of Solar Electricity*, Aatec Publications (a:3-4, b:16-20, c:25, d:viii, e:42, f:46, g:11)
88. "applications of energy-related materials" from *Encyclopædia Britannica*, 2004
89. "silicon" from *Encyclopædia Britannica*, 2004
90. J. Diamond, 1997, *Guns, Germs, and Steel: The Fates of Human Societies*, W.W. Norton & Company (a:89)
91. F. Fukuyama, 2002, *Our Posthuman Future: Consequences of the Biotechnology Revolution*, Farrar Straus and Giroux
92. F.J. Dyson, 1999, *The Sun, The Genome, The Internet*, Oxford University Press
93. R. Chau et al., 2003, *Silicon nano-transistors for logic applications*, Physica E, 19 (1-5)
94. L.R. Brown, C. Flavin and H. French, 1999, *State of the World 1999*, The Worldwatch Institie, W.W. Norton & Company (a:5-10, b:41-46, c:136)
95. J. van Klinken, 2003, *Energy Constraints with Limited Options*, in [119] (a:60-1)
96. World Wide Fund for Nature, 2002, *Living Planet Report 2002* (a:22, b:2)
97. Y. Baskin, 1997, *The Work of Nature: How the Diversity of Life Sustains Us*, Island Press (a:155)
98. World Resources Institute, United Nations Environment Programme and the World Business Council for Sustainable Development, 2002, *Tomorrow's Markets: Global Trends and Their Implications for Business* (a:52, b:26-7)
99. K.S. McCann, 2000, *The diversity-stability debate*, Nature, 405 (228-233)
100. P. Atkins, 1984/1994, *The 2nd Law: Energy, Chaos, and Form*, Scientific American Library (a:8-9, b:62-3, c:183-6)
101. Q. Schiermeier, 2003, *Setting the record straight*, Nature, 424 (482-3)
102. L.P. Koh, R.R. Dunn, N.S. Sodhi, R.K. Colwell, H.C. Proctor and V.S. Smith, 2004, *Species Coextinctions and the Biodiversity Crisis*, Science, 305 (1632-4)

References

103. R. Dawkins, 1993, *The Evolutionary Future of Man: A Biological View of Progress*, The Economist, 328 (87)
104. S.E. Jørgensen, 2001, *Thermodynamics and Ecological Modelling*, CRC Press LLC (a:157)
105. "Smith, Adam" from *Encyclopædia Britannica*, 2004
106. É. Danchin, L.A. Giraldeau, T.J. Valone and R.H. Wagner, 2004, *Public Information: From Nosy Neighbors to Cultural Evolution*, Science, 305 (487-91)
107. R. Dawkins, 1976/1989, *The Selfish Gene*, 2nd edition, Oxford University Press (a:192)
108. R. Dawkins, 1982/1999, *The Extended Phenotype*, revised edition with new Afterword and Further Reading 1999, Oxford University Press (a:109)
109. R. Aunger, 2002, *The Electric Meme: A New Theory of How We Think*, The Free Press (a:197-9, b:181, c:139, d:314-5, e:30, f:227, g:245-6, h:291, i:298, j:301, k:304, l:277)
110. "noösphere" from *Encyclopædia Britannica*, 2004
111. J. Verloop, 2004, *Insight in Innovation: Managing Innovation by Understanding the Laws of Innovation*, Elsevier (a:6, b:12-6)
112. E.S. Cassedy, 2000, *Prospects for Sustainable Energy: a Critical Assessment*, Cambridge University Press (a:232)
113. S. Blackmore, 1999, *The Meme Machine*, Oxford University Press (a:40)
114. N. Valéry, 1999, *Innovation in Industry*, The Economist, February 20th
115. N. McDowell, 2002, *Ecological footprint forecasts face skeptical challenge*, Nature, 419 (656)
116. World Wide Fund for Nature (WWF), 2002, *Living Planet Report 2002*
117. W.E. Rees, 2003, *A blot on the land*, Nature, 421 (898)
118. J.A. Thomas, M.G. Telfer, D.B. Roy, C.D. Preston, J.J.D. Greenwood, J. Asher, R.Fox, R.T. Clarke and J.H. Lawton, 2004, *Comparative Losses of British Butterflies, Birds, and Plants and the Global Extinction Crisis*, Science, 303 (1879-81)
119. B. van der Zwaan and A. Petersen (eds), 2003, *Sharing the Planet: Population – Consumption – Species*, Eburon Publishers
120. B. van der Zwaan and A. Petersen, 2003, *The Precautionary Principle: (Un)certainties about Species Loss*, in [119] (a:141)
121. International Institute for Sustainable Development (IISD), 1999, *Sustainable Development Timeline*, 2nd edition
122. J. de Rosnay, 1975/1979, *The Macroscope: A New World Scientific System*, Harper & Row, Publishers, New York; First published in France under the title *Le Macroscope: Vers une Vision Globale*, Editions du Seuil, 1975
123. H.-J. Schellnhuber, 1999, *'Earth system' analysis: and the second Copernican revolution*, Nature, 402 (C19-C23)
124. D. Worster, 1977/1994, *Nature's Economy: A History of Ecological Ideas*, 2nd edition, Cambridge University Press (a:2, b:30, c:22-3, d:53)
125. H. Achterhuis, 1998, *De Erfenis van de Utopie*, Ambo (a:192)
126. S. Dresner, 2002, *The Principles of Sustainability*, Earthscan Publications Ltd (a:101)
127. R. Douthwaite, 1999, *The Growth Illu$ion: How Economic Growth has Enriched the few, Impoverished the Many and Endangered the Planet*, New Society Publishers
128. G.G. Marten, 2001, *Human Ecology: Basic Concepts for Sustainable Development*, Earthscan Publications Ltd (a:175, b:17-20)
129. M. Kaku, 1997, *Visions: How Science will Revolutionize the 21st Century*, Anchor Books (a:5, b:116, c:9-10)
130. P. Menzel and F. D'Alusio, 2000, *Robo sapiens: Evolution of a New Species*, The MIT Press (a:16)
131. H. L. Sweeney, 2004, *Gene Doping*, Scientific American, July (64)
132. W. W. Gibbs, 2004, *Synthetic Life*, Scientific American, May (75)

133. B. Lomborg, 2001, *The Skeptical Environmentalist: Measuring the Real State of the World*, Cambridge University Press (a:323)
134. V. Smil, 2003, *Energy at the Crossroads: Global Perspectives and Uncertainties*, The MIT Press (a:183, b:204, c:218-21, d:215, e:309, f:313, g:133)
135. United Nations Environment Programme (UNEP) and International Energy Agency (IEA), 2002, *Reforming Energy Subsidies* (a:12)
136. International Energy Agency (IEA), 2003, *Energy technology: facing the climate challenge* (a:3)
137. European Community, 2004, *Nuclear Energy*, RTD info No 40 (a:12)
138. F. Laroui and B.C.C. van der Zwaan, 2002, *Environment and Multidisciplinarity: Three Examples of Avoidable Confusion*, Integrated Assessment, 3(4) (360-369)
139. M.H. Key, 2001, *Fast track to fusion energy*, Nature, 412 (775-6)
140. International Energy Agency (IEA), 2003, *Technology Options: Fusion Power*, Energy Technology Policy & Collaboration Papers
141. S. Abraham, 2004, *The Bush Administation's Approach to Climate Change*, Science, 305 (616-7)
142. International Energy Agency (IEA), 2004, *Key World Energy Statistics 2004*
143. World Business Council for Sustainable Development (WBCSD), 2004, *Facts and trends to 2050: Energy and climate change*
144. L. Reijnders, 2003, *Loss of Living Nature and Possibilities for the Limitation Thereof*, in [119] (a:126)
145. E. von Weizsäcker, A.B. Lovins and L.H. Lovins, 1998, *Factor Four: Doubling Wealth, Halving Resource Use*, Earthscan Publications Ltd
146. P. Hawken, A.B. Lovins and L.H. Lovins, 1999, *Natural Capitalism: The Next Industrial Revolution*, Earthscan Publications Ltd
147. E.F. Schumacher, 1973/1993, *Small is Beautiful: a Study of Economics if People Mattered*, Vintage Books
148. United Nations Department of Economic and Social Affairs, Population Division, 2004, *World Population to 2300*, United Nations
149. M. ElBaradei, 2004, *Nuclear Power: An Evolving Scenario*, IAEA Bulletin 46/1, International Atomic Energy Agency
150. E. von Weizsäcker, 2003, *From 'Limits to Growth' to 'Factor Four'*, in [119] (a:15)
151. E.S. Goodstein, 1999, *Economics and the Environment*, 2nd edition, Prentice-Hall, Inc. (a:81-2)
152. K. van der Wal, 2003, *Globalisation, Environment and Ethics: the Problematic Relationship between Modernity and Sustainability*, in [119] (a:173)
153. J. van den Bergh, 1999/2002, *Handbook of Environmental and Resource Economics*, Edward Elgar
154. J. van den Bergh and R. de Mooij, 1999/2002, *An assessment of the growth debate*, in [153] (a:653, b:648)
155. S. Baumgärtner, 2003, *Entropy*, Internet Encyclopaedia of Ecological Economics, International Society for Ecological Economics
156. N. Georgescu-Roegen, 1971/1999, *The Entropy Law and the Economic Process*, Harvard University Press (a:304, b:306)
157. H.E. Daly, 1996, *Beyond Growth: The Economics of Sustainable Development*, Beacon Press (a:31-3, b:49)
158. Merriam-Webster Online Dictionary, 2004

References

159. R. Dawkins, 2001, *Sustainability doesn't come Naturally: a Darwinian Perspective on Values*, The Environment Foundation, The Values Platform for Sustainability, Inaugural lecture, Wednesday, 14 November 2001, the Royal Institution
160. R. Gerlagh and B. van der Zwaan, 2002, *Long-Term Substitutability between Environmental and Man-Made Goods*, Journal of Environmental Economics and Management, 44 (329-345)
161. B. de Gaay Fortman, 2003, *Environmental Security Revisited: A Civilisational Perspective*, in [119] (a:197)
162. H. Gosselink, 2002, *Pathways to a more Sustainable Production of Energy: Sustainable Hydrogen – a Research Objective for Shell*, International Journal of Hydrogen Energy, 27 (1125-9)
163. K. Humphreys and M. Mahasenan, 2002, *Toward a Sustainable Cement Industry – Substudy 8: Climate Change*, An independent study commissioned by the World Business Council for Sustainable Development (WBCSD) by Battelle (a:10, b:28, c:v)
164. I. Prigogine, 1996/1997, *The End of Certainty: Time, Chaos, and the New Laws of Nature*, The Free Press (a:66-7, b:158, c:202)
165. D. Kondepudi and I. Prigogine, 1998, *Modern Thermodynamics: From Heat Engines to Dissipative Structures*, John Wiley & Sons (a:431-2, b:427, c:452, d:5-6, e:84)
166. "Helmholtz, Hermann von" from *Encyclopædia Britannica*, 2003
167. J. Kay, 2000, *Ecosystems as Self-Organizing Holarchic Open Systems: Narratives and the Second Law of Thermodynamics*, in *Handbook of Ecosystem Theories and Management*, ed. by S.E. Jørgensen, CRC Press, Lewis Publishers (135-160)
168. E. Scheider and J. Kay, 1995, *Order from Disorder: The Thermodynamics of Complexity in Biology*, in *What is Life: The Next Fifty Years: Reflections on the Future of Biology*, ed. by M. Murphy and L. O'Neill, Cambridge University Press (161-72)
169. J. Kay, T. Allen, R. Fraser, J. Luvall and R. Ulanowicz, 2000, *Can we use energy based indicators to characterize and measure the status of ecosystems, human, disturbed, and natural?*, in Proceedings of the International Workshop: Advances in Energy Studies: Exploring Supplies, Constraints and Strategies, Porto Venere, Italy, 13–17 May (121-33)
170. E. Scheider and J. Kay, 1994, *Life as Manifestation of the Second Law of Thermodynamics*, Mathematical and Computer Modelling, 19 (6-8) (25-48)
171. P. Bunyard (ed.), 1996, *Gaia in Action: Science of the Living Earth*, Floris Books (a:15)

Glossary

Acheulean industry Stone-tool industry of the Lower Paleolithic Period characterised by bifacial stone tools with round cutting edges and typified especially by an almond-shaped flint hand axe measuring 20 to 25 cm in length and flaked over its entire surface. Other implements include cleavers, borers, knives, and choppers. The name derives from the site near St.-Acheul, in northern France, where such tools were first discovered.* The Acheulean industry persisted largely unchanged for about one million years after its origination about 1.7 million years ago, when it replaced the Oldowan industry [54h,i].

Adenosine triphosphate (ATP) Organic compound, substrate in many enzyme-catalysed reactions in the cells of animals, plants, and microorganisms. As ATP's chemical bonds store a substantial amount of chemical energy, it could function as the carrier of chemical energy from energy-yielding oxidation of food to energy-demanding cellular processes.* ATP is the principal energy currency of intracellular energy transfer [33a] or the universal energy transmitter in biological systems [34a]. Fermentation and cellular respiration are metabolic processes that are sources of ATP.

Aerobe An organism that lives and grows in the presence of oxygen.

Aerobic respiration See Respiration.

Age of Reason See Enlightenment.

Agricultural Regime Energy regime during the Agrian Period characterised by agricultural technology.

Anaerobe An organism that lives and grows in the absence of oxygen.

Anaerobic respiration See Respiration.

Anthropic Era The period covering the evolution of the anthroposphere, which began circa 2.5 million years ago with the origin of genus *Homo*.

Anthropocentrism Regarding humankind as the centre of existence.

Anthropogenic Generated by human beings.

Anthropomorphism The attribution of human characteristics to other beings or objects.

Anthroposphere (or humansphere) Subsystem of 'system Earth' made up of the artosphere and the noösphere.

Anthroposystem Subsystem of the anthroposphere. Four anthroposystems can be distinguished: human knowing, human capacity, human acting and human co-existing or living together.

---

*(from *Britannica Concise Encyclopædia*, 2005).

Arcadian tradition The 'Arcadian' stance towards nature advocates a simple, humble life for man, with the aim of restoring him to a peaceful coexistence with other organisms [124a]. In our time *Homo ecologicus* has come to represent the Arcadian mood to austerely conserve nature. (See also Imperial tradition.)

Archaean The second aeon on the geological time scale of the history of the Earth, which began with the formation of the Earth's crust about 4.0 billion years ago and lasted until about 2.5 billion years ago [47b].

Artosphere The whole of human artefacts; 'arto' refers to human-made. The term 'technosphere' is more common, but not all artefacts are technological by nature.

Attractor state (also dynamic steady state) A state adopted by an energy-dissipating structure that shows the tendency to counteract perturbations enforced from outside.

Autocatalysis The product of a chemical reaction which acts as a catalyst for its own production.

Autotrophy When a living organism needs only carbon dioxide as carbon source for its 'biosynthetic works' [34b]. Autotrophs stand at the beginning of food chains and are also called primary producers, because the organic molecules they make serve as food for other organisms, called heterotrophs [39a].

Biogenic Generated by biological life.

Biosphere See Ecosphere.

Blue-greens See Cyanobacteria.

Cambrian Period The earliest time division of the Paleozoic Era, extending from about 540 to 505 million years ago.[†]

Carbocultural Regime Energy regime during the Carbian Period characterised by the exploitation and use of fossil fuels.

Carbohydrate Any member of the class of organic compounds composed of carbon, hydrogen and oxygen including sugars, starch, and cellulose, commonly classified as mono-saccharides (or simple sugars such as glucose and fructose), disaccharides (or two-unit sugars such as sucrose and lactose), oligosaccharides (containing 3 to 10 or so sugars), and polysaccharides (large molecules with up to 10,000 monosaccharide units, including cellulose and starch), as well as functionalised derivatives.[*] Photosynthetic organisms such as green plants produce carbohydrates from carbon dioxide and water.

Carbon Chemical element. The most common stable isotope is carbon-12. Carbon-14 is the best known of five radioactive isotopes; its half-life of approximately 5730 years makes it useful in carbon-14 dating. Pure carbon occurs for example in diamond and graphite. With hydrogen, oxygen, nitrogen and a few other elements, carbon forms the compounds that make up all living things: proteins, carbohydrates, lipids, and nucleic acids. Carbon dioxide makes up about 0.03 percent of the air, while carbon occurs in the Earth's crust as carbonate rocks and hydrocarbons in the form of fossil minerals such as coal, petroleum, and natural gas. The oceans contain large amounts of dissolved carbon dioxide and carbonates.[*]

---

[*](from *Britannica Concise Encyclopædia*, 2005).

[†](from *Encyclopædia Britannica*, 2003–5).

Glossary

Carbon dioxide A colourless gaseous product that results from the complete combustion of organic compounds such as fossil fuels. Carbon dioxide is a main greenhouse gas.

Carrying capacity Maximum population of a particular species that a given habitat can support over a given period of time [39a].

Catalysis (in chemistry) Modification of the rate of a chemical reaction, usually an acceleration, by a substance, known as a catalyst, which is not consumed during the reaction.[†]

Cenozoic Era Third of the major eras of the Earth's history, and the interval of time during which the continents assumed their current geographic positions. It was also the time when the Earth's flora and fauna evolved towards those of the present day. The Cenozoic, from the Greek for 'recent life', began circa 65 million years ago and is divided into two periods, the Tertiary (65 to 1.8 million years ago) and the Quarternary (1.8 million years ago to the present).[*]

Chemotrophy When a living organim uses oxidisable compounds from the environment as energy source [39c]; compare phototrophy.

Chlorophyll Any member of a class of light-sensitive pigments involved in photosynthesis.[†]

Chloroplast Microscopic, ellipsoidal organelle in a green plant cell. It is the site of photosynthesis, distinguished by its green colour caused by the presence of chlorophyll.

Clean fossil The use of fossil fuel with practically no emissions, including CO_2, into the environment.

Co-evolution The parallel evolution of two kinds of living organisms, like flowers and their pollinators, or living organisations, such as eco- and/or sociosystems, which are interdependent and where any change in one will induce an adaptive response in the other [6a].

Co-generation Simultaneous production of two useful forms of energy, such as heat and electricity, from the same fuel source [39a]; compare Combined heat and power.

Combined heat and power (CHP) Production of heat and electricity from one fuel source such as natural gas; compare Co-generation.

Complex adaptive system (CAS) A free-living energy-dissipating structure [14f,g]. Basal information units of a CAS, such as a gene, or a meme, are sensitive to mutagenesis.

Complexity A measure of the total of information needed to describe a system's structure and function [36b].

Complexity, societal Refers to the number of social personalities within a society. Hunter-gatherer societies contain no more than a few dozen distinct social personalities, while modern European censuses recognise 10,000 to 20,000 unique occupational roles, and industrial societies may contain overall more than 1,000,000 different kinds of social personalities [72].

Consilience (by Edward O. Wilson) Applies if explanations of different phenomena can be connected and proved consistent with one another [69c].

---

[*](from *Britannica Concise Encyclopædia*, 2005).

[†](from *Encyclopædia Britannica*, 2003–5).

Cyanobacteria (or blue-greens) Any of a large group of prokaryotic, mostly photosynthetic organisms. They contain certain pigments, which, with their chlorophyll, often give them a blue-green colour (though many species are actually green, brown, yellow, black, or red). Cyanobacteria played a crucial role in raising the level of free oxygen in the atmosphere of the early Earth.*

Cytoplasm Part of a eukaryotic cell outside the nucleus. It contains all the organelles, including the mitochondria and chloroplasts, the cytoskeleton and the cytosol (the fluid mass that surrounds the various organelles).*

Cytoskeleton A system of tiny filaments or fibres that is present in the cytoplasm of eukaryotic cells. It organises other constituents of the cell, maintains the cell's shape, and is responsible for the locomotion of the cell itself and the movement of the various organelles within it.†

Cytosol See Cytoplasm.

Deuterium (or heavy hydrogen) Isotope of hydrogen, chemical symbol ^{2}H or D. Its nucleus contains one proton and one neutron, whereas the nucleus of normal hydrogen comprises only a proton. Deuterium is found in naturally occurring hydrogen compounds to the extent of about 0.015 percent. Analogous to ordinary water, or H_2O, it forms heavy water, or D_2O.*

Diffusion (in economics) The spread of a new product through a community, which enables the development of new *ways of living*.

Discovery As the word says, a dis ... covery, or an act of removing the 'cover' from *something that exists*. It is a first-time *observation* of an object, a phenomenon, a pattern, or a quality of natural or cultural origin.

Dynamic steady state (also attractor state) A non-equilibrium condition whereby the thermodynamic properties of an open system vary, but within certain boundaries.

Ecocentrism The perception of reality that all that Nature provides has intrinsic value [152a].

Ecological economics The view that natural and economic capital are complementary; that is, they are used together in production having low substitutability. Natural systems are viewed as rather fragile, and if one component is disturbed, the productivity of the entire ecosystem may collapse [151a].

Ecology Study of the interactions of living organisms with one another and their non-living environment of energy and matter; study of the structure and functions of nature [39a].

Ecosphere The sum of the Earth's ecosystems, or the worldwide collection of living organisms interacting with one another and their non-living environment of energy and matter. Also called the biosphere [39a].

Ecosystem The complex of living organisms, their physical environment, and all their interrelationships in a particular unit of space.†

---

*(from *Britannica Concise Encyclopædia*, 2005).

†(from *Encyclopædia Britannica*, 2003–5).

Glossary

Emergence (in energetics and complexity science) The spontaneous emergence of structure in an energy-dissipating or complex adaptive system. Each energy-dissipating structure has the intrinsic potential to develop new structure spontaneously. A whirlwind is a physical emergent property of hot air; a species is a biological emergent property of the biosphere; a population distribution is an ecological emergent property of an ecosystem; and, Adam Smith's 'invisible hand' is a societal emergent property of a human economy.

Endosymbiosis See Symbiogenesis.

Energetics The science that studies energy and its transformations.

Energy-dissipating structure Spatiotemporal structure that appears in a far-from-equilibrium condition [164c].

Energy gradient Entropic difference between an energy source and an energy sink.

Energy period A time-span characterised by an ecologically dominant energy regime.

Energy regime An eco- or sociosystem characterised by an energy system, or energy economy, or a network of distinct energy chains.

Energy sink A relatively high entropy state.

Energy source A relatively low entropy state.

Enlightenment (also Age of Reason) European intellectual movement of the 17th to 18th century in which ideas concerning God, reason, nature, and man were blended into a worldview that inspired revolutionary developments in art, philosophy, and politics. Central to Enlightenment thought were the use and celebration of reason. In the sciences and mathematics, the logics of induction and deduction made possible the creation of a sweeping new cosmology. One of the Enlightenment's enduring legacies is the belief that human history is a record of general progress.*

Entropy A measure of the disorder or randomness of a system: the greater the disorder of a system, the higher its entropy; the higher its order, the lower its entropy [39a]. For an isolated system the entropy function reaches its maximum at thermodynamic equilibrium [164c].

Equilibrium See Thermodynamic equilibrium.

Eukaryote Any organism composed of one or more cells, each of which contains a clearly defined nucleus enclosed by a membrane, along with organelles (small, self-contained, cellular parts that perform specific functions) such as mitochondria and chloroplasts. Note that not all eukaryotes carry chloroplasts; only the photosynthesising ones.*

Evolution (in biology) Theory postulating that the various types of animals and plants have their origin in other pre-existing types and that the distinguishable differences are due to modifications in successive generations† originated in a process of descent with modification.

Evolution (in energetics) The development in time of energy regimes characterised by distinct energy-dissipating structures that could be biogenic or anthropogenic in

---

*(from *Britannica Concise Encyclopædia*, 2005).

†(from *Encyclopædia Britannica*, 2003–5).

nature. As energy regimes are historically connected and their development proceeds through a process of descent with modification, their evolution is Darwinian by nature.

Evolutionary energetics The branch of energetics developing evolutionary theories of energy and its transformations; just as evolutionary biology is a branch of biology; and evolutionary economics is a branch of economics.

Feedback Occurs when a change in a variable triggers a response that also affects the variable itself; a 'negative' feedback then tends to damp the initial change whereas a 'positive' tends to amplify it [47b].

Fermentation See Respiration.

Fertile Crescent The term describes a crescent-shaped area of arable land, probably more agriculturally productive in antiquity than it is today. Historically, the area stretched from the south-eastern coast of the Mediterranean Sea around the Syrian Desert north of the Arabian Peninsula to the Persian Gulf; in general, it often includes the Nile River valley as well.*

Fire Regime Same as Pyrocultural Regime.

Force, driving (this work) A creative force that results from an energy flow through an energy-dissipating structure. For example, gravity is a driving force in the emergence of a whirlpool.

Force, shaping (this work) A structuring force that results from interactions between the composing parts of an energy-dissipating structure. For example, the physico-chemical interactions between water molecules are a shaping force in the structuring of a whirlpool.

Footprint, ecological A measure of man's use of renewable natural resources [96b] in terms of *"the area of productive land and water that people need to support their consumption and to dispose of waste"* [115]. In other words, it quantifies the total area required to produce the food and fibres that a country consumes to sustain its energy consumption, and to give space for its infrastructure [116].

Gaia theory Theory of the Earth as a system in which the evolution of organisms and environment are tightly coupled, and self-regulation of climate and chemical composition are emergent properties [47b].

Gene Unit of biological hereditary information situated on a particular locus on a chromosome, composed of DNA and proteins [6a].

Genetics Branch of biology dealing with evolutionary change in populations, or the heredity and variation of organisms.

Genome A collective term for the inheritable, or genetic material of an organism.

Genotype The genetic makeup of an organism.*

Global warming See 'Greenhouse effect, enhanced'.

Green energy A synonym for renewable energy.

---

*(from *Britannica Concise Encyclopædia*, 2005).

Glossary

Greenhouse effect, enhanced (or global warming) Greenhouse effect by anthropogenic emissions, such as aerosols, methane, and carbon dioxide, enhancing the natural greenhouse effect.

Greenhouse effect, natural Greenhouse gases maintain the Earth's temperature about 33°C higher than it would be in the absence of them [39f] – which probably would render the planet too cold to nourish life.

Greenhouse gas Component of the atmosphere which contributes to the retention of energy received from solar radiation by absorbing re-radiated energy from the Earth's surface [47b].

Hadean The first great division, or aeon, of the geological time scale of the Earth's history, from about 4.55 to 4.0 billion years ago [47b].

Heterotrophy When a living organism needs organic compounds as carbon source for its 'biosynthetic works' [34b]. Except for some prokarya, heterotrophs are chemotrophs (including animals and humans): they feed wholly on the tissues of autotrophs (in case of herbivores, or primary consumers), or of other heterotrophs (in case of carnivores and omnivores), supplying both energy and organic building blocks [39a]; also, carnivores consume energy-rich biomatter that ultimately comes from autotrophs eaten by the prey.

Holism The philosophy that the universe and especially living nature is seen in terms of interacting wholes that are more than the mere sum of elementary particles [158].

Homeostasis Any self-regulating process by which a biological or mechanical system maintains stability while adjusting to changing external conditions.*

Hominid Any creature of the family Hominidae.*

Hominidae The taxonomic family (order Primates) that includes modern humans and their direct extinct ancestors, or the human line. *Homo sapiens* is the only species in the genus *Homo* of the family Hominidae living today.†

Hydrothermal system Circulation system of water in fractures and pores around a hot body of rock [47b].

Hydrocarbon Any of a class of organic compounds composed only of the chemical elements carbon and hydrogen. The carbon atoms form the framework, whereas the hydrogen atoms attach to them. Hydrocarbons are the principal constituents of fossil fuels.

Hyperthermophile An organism having an optimal growth temperature of 80°C or higher [47b].

Imperial tradition The 'Imperial' stance toward nature aspires to establish, through the exercise of reason and hard work, man's dominion over nature [124a] *Homo economicus* has come to represent the imperial mood in following founding genius Adam Smith, who *"saw nature as no more than a storehouse of raw materials for man's ingenuity"* [124d] (see Neoclassical economics; see also Arcadian tradition).

---

*(from *Britannica Concise Encyclopædia*, 2005).

†(from *Encyclopædia Britannica*, 2003–5).

Industrial Revolution The process of change from an agrarian, handicraft economy to one dominated by industry and machine manufacture originating in the late 18th-century in England [76].

Innovation The economically successful, practical application of a newly invented tool or method.

International Energy Agency (IEA) Established in November 1974 in response to the oil crisis as an autonomous intergovernmental entity within the Organisation for Economic Cooperation and Development (OECD) to ensure the energy security of industrialised nations.

Invention The creation of a new capacity based on new knowledge. An invention is something that did not exist before and enables the development of applications in the form of new tools or techniques.

Isotope One of two or more species of atoms of a chemical element having nuclei with the same number of protons but different numbers of neutrons.*

Levantine Corridor The narrow band of habitable land in Israel and Jordan that offered the most obvious way for the ancestors of modern humans to disperse out of Africa [65].

Lignin Complex oxygen-containing organic compound, a mixture of polymers of poorly known structure. After cellulose, it is the most abundant organic material on Earth making up one-fourth to one-third of the dry weight of wood where it is concentrated in the cell walls.*

Macroscope (from *macro*, great, and *skopein*, to observe) A symbolic instrument composed of scientific methods and modern technologies from different disciplines [122]. It is designed with a view to observing connections in complex adaptive systems such as eco- and sociosystems up to 'system Earth'.

Macroscopical Signal The contemporary surge of macroscopical observations that possibly indicates the emergence of a new dominating perception of reality.

Meme The information unit of cultural inheritance, coined by Richard Dawkins. A meme should be regarded as a unit of information that has a definite structure, physically residing in the brain. Its consequences in the outside world are phenotypic effects, *cf.* genes [108a].

Memetics Study of developmental processes in terms of the evolution of memes.

Mesozoic Era Second of the Earth's three major geologic eras and the interval during which the continental landmasses as known today were separated from the supercontinents by continental drift. It lasted from circa 248 to circa 65 million years ago.*

Metabolism (in biology) The sum of the physicochemical conversions that take place within each cell of a living organism to provide energy and matter for all vital, biogenic processes.

Metazoa A subkingdom comprising all multicellular animals [47b].

Metazoan Multicellular organism whose cells are differentiated into tissues and organs, or multicellular animal [50].

---

*(from *Britannica Concise Encyclopædia*, 2005).

Glossary

Methanotroph An organism capable of oxidising methane [47b].

Microprocessor Miniature electronic device that can interpret and execute software program instructions as well as handle arithmetic operations. In the late 1970s the microprocessor enabled the development of the microcomputer by becoming its central processing unit, or CPU.*

Mitochondrion A eukaryotic organelle, aptly called a cellular powerhouse. Mitochondria are responsible for most of the cell's respiration and energy production in the form of adenosine triphosphate (ATP).†

Motility From 'motile', that is exhibiting or capable of movement [158].

Mutagen (in biology) Any agent capable of altering a cell's genetic makeup by changing the structure of the hereditary material, DNA. Many forms of electromagnetic radiation are mutagenic, as are various chemical compounds.

Mutagenesis The emergence of a mutant through mutation, that is alteration of information encoded by a unit of inheritance, such as a gene or a meme.

Mutant A viable mutation of a free-living energy-dissipating structure, or complex adaptive system.

Nanometre A billionth of a metre.

Nanotechnology An all-embracing concept referring to the technology needed to manipulate, modify or mold matter on the nano-scale; genetic engineering and modern digital computing are manifestations of the nano-technological advancement.

Naturalism Branch of biology dealing with evolutionary change in biodiversity, and particularly the origin of new species and higher taxa [4].

Natural selection The process by which, in every generation, individuals of lower fitness are removed from the population [6a].

Neoclassical economics The view that natural and man-made capital are substitutes in production: as resources become scarce, prices will rise, and human innovation will yield high-quality substitutes, lowering prices once again. Nature is seen as highly resilient; pressures on ecosystems will lead to steady, predictable degradation. This perception of smooth substitution between inputs and small natural changes is at the heart of the neoclassical tradition, in addition to the belief in the power of the market-based economy to improve the quality of life [151a].

Neolithic Period (or New Stone Age) Final stage of technological development or cultural evolution among prehistoric humans; followed the Palaeolithic Period, or Old Stone Age.*

Noösphere The Greek word 'noös' means 'mind', so noösphere stands for 'mindsphere' [110], or the sphere of thoughts created and shared by human beings.

Oldowan industry Stone-tool industry of the early Paleolithic (between circa 2.5 and 1.6 million years ago [54i]) characterised by crudely worked pebble tools. Oldowan tools,

---

*(from *Britannica Concise Encyclopædia*, 2005).

†(from *Encyclopædia Britannica*, 2003–5).

made of quartz, quartzite, or basalt, are chipped in two directions to form simple, rough implements for chopping, scraping, or cutting. The industry is associated with early hominids and has been found at Olduvai Gorge (from which its name derives), Lake Turkana, and the Afar region of Ethiopia.*

Organelles See Eukaryote.

Oxygenesis (or oxygenic photosynthesis) The biological process in which organic compounds are synthesised from carbon dioxide and water using the free energy from sunlight, and liberating oxygen. For example, cyanobacteria (blue-greens), algae and plants perform oxygenic photosynthesis [47b].

Palaeolithic Period (or Old Stone Age) Ancient technological or cultural stage characterised by the use of rudimentary chipped stone tools. Three major subdivisions – Lower, Middle, and Upper Paleolithic – are recognised in Europe, although the dividing line between the Lower and Middle stages is not so clearly defined as that separating the Middle and Upper subdivisions. Lower Paleolithic lasted from circa 2.5 million years to 200,000 years ago; Upper Paleolithic from 40,000 to 10,000 BC.†

Paleo-record The whole of observable physical data related to ancient forms of life and corresponding ecological conditions contained by planet Earth.

Paleozoic Era Major interval of geologic time, circa 543 to 248 million years ago.*

Pastoralism, nomadic Early agriculture based on raising livestock as the primary economic activity [158].

Permian Period Interval of geologic time, 290 to 248 million years ago. The last of the six periods of the Paleozoic Era.*

Phenotype All the observable characteristics of an organism, such as shape, size, colour, and behaviour, that result from its interaction with the environment.*

Photon Light quantum, or discrete energy packet in light. The energies of photons range from high-energy gamma rays and X rays to low-energy infrared and radio waves, though all travel at the same speed, the speed of light. Photons have no electric charge or rest mass.*

Photosynthesis The biological process in which light energy is converted into chemical energy, then used for the production of organic compounds from carbon dioxide through reduction [47b]. In oxygenic photosynthesis water is the reductant. Anoxygenic photosynthesis involves other reductants, such as hydrogen sulphide.

Photosphere Visible surface of the Sun, about 400 km thick, which emits the sunlight that reaches the Earth.†

Phototrophy When a living organism uses light as energy source [39c]; compare chemotrophy.

Photovoltaic cell (or solar cell) Device that directly converts solar energy into electrical energy.

---

*(from *Britannica Concise Encyclopædia*, 2005).

†(from *Encyclopædia Britannica*, 2003–5).

Glossary

Prokaryote An organism lacking a nucleus and other membrane enclosed organelles, and usually having its DNA as a single stranded circular molecule [47b].

Proterozoic The third aeon on the geological time scale of the Earth's history, from about 2.5 billion to 0.55 billion years ago.

Protozoan Single-celled eukaryotic organism including the most primitive forms of life [47b].

Pyrocultural Regime Energy regime during the Pyrian Period characterised by anthropogenic fire control.

Resilience The ability of an eco- or sociosystem to simultaneously balance operational stability and the power to adapt to external change.

Quantificational Signal A marked shift in people's 'quantification attitude' that surged in the period between 1250 and 1350. People began to think in quanta time, space, and value: the mechanical clock, marine charts, perspective painting, double-entry bookkeeping, and note length in musical notation, came to prominence. Also money as an abstract measurement of 'worth' appeared [16b]. In effect the Quantificational Signal deeply touched human's perception of reality. It initialised revolutionary change, first the Scientific, and later the Carbocultural Revolution.

Reductionism The study of the world as an assemblage of physical parts that can be broken apart and analysed separately [69a].

Respiration The process by which living organisms gain energy, and make adenosine triphosphate (ATP), through oxidation of organic substrates. If the oxidant is oxygen, respiration is aerobic; anaerobic respiration utilises oxidants from the environment other than oxygen; a special form of anaerobic respiration is fermentation, which applies internally generated electron acceptors [34d].

Science schism The break in academic unity that emerged in the 19th century and divides the natural sciences and the cultural branches of *arts* and *humanoria*.

Sedentism Living permanently in one place, a lifestyle that emerged with agriculture.

Self-organisation A term used to characterise the phenomenon of spontaneous emergence (see Emergence).

Shifting cultivation Early agriculture based on raising crops as the primary economic activity. It involved alternation between short periods of cropping – commonly just one season; rarely more than three years – and long spans of fallow, lasting at least a decade. The cropping cycle often began with partial removal of natural overgrowth: accomplished in forests by 'slash and burn' and on grasslands simply by setting fires [33f].

Socio-metabolism (or societal metabolism) The whole of the energy and materials flows going through human communities, or the sum of the physico-chemical conversions that take place within a human society providing energy and matter for all economic and non-economic anthropogenic processes (see Metabolism).

Steady state A non-equilibrium state wherein the thermodynamic properties of an open system are constant in time; thus energy, or energy and mass, transfer(s) across system boundaries.

Symbiosis (in biology) Any of several prolonged living arrangements, or physical associations, between members of two or more different species [7p]. The members are called symbionts.

Symbiosis, memetic (this work) A prolonged cultural arrangement evolving its own way (as between the Imperial and Arcadian worldviews) [158].

Symbiogenesis The origin of new tissues, organs, organisms – even species – by establishment of long-term or permanent symbiosis [51b]. Eukaryotes could originate through microbial symbiogenesis, or acquisition and inheritance of cellular symbionts [7f]. The term endosymbiosis is also used to indicate emergence of cellular partnerships with one cell nested inside the other [43j].

Symbolisational Signal The uniquely human capacity to symbolise that evolved in social groups of fire masters. The Symbolisational Signal is revealed by the archaeological annals exposing a sudden growth of symbolic activity with the arrival of *Homo sapiens* 50,000 to 40,000 years ago. It indicates that human beings adopted a new perception of reality allowing them to think abstractly about Nature and themselves.

Taxon Any unit used in taxonomy; taxa are arranged in a hierarchy from kingdom to subspecies, a given taxon ordinarily including several taxa of lower rank.[†]

Taxonomy The science of biological classification.[†]

Technosphere See Artosphere.

Thermodynamic equilibrium If a physical system is isolated, its state – specified by variables such as pressure, temperature, and chemical composition – evolves irreversibly towards a time-invariant condition in which no physical or chemical change in the system occurs spontaneously. This is the state of 'thermodynamic equilibrium', characterised, among others, by a uniform temperature throughout the system [165d].

Thermodynamics The study of the transformations of energy in all its forms [100a]. The First Law of Thermodynamics reads that energy is conserved. The Second Law of Thermodynamics says that the sum of the entropy changes of a system and its exterior can never decrease [165e]; in other words, the Second Law holds that natural, and thus spontaneous, change accompanies the fall of energy from a higher to a lower quality, or rather from a lower to a higher entropy.

Tritium Isotope of hydrogen, chemical symbol ^3H or T. Its nucleus contains one proton and two neutrons (normal H has none, deuterium one neutron). Tritium is radioactive, with a half-life of 12.32 years. It rarely occurs naturally: in water the amount is 10^{-18} that of ordinary hydrogen.*

Vertebrate An animal with an internal skeleton.[†]

Vortex Something that resembles a whirlpool.

---

*(from *Britannica Concise Encyclopædia*, 2005).

[†](from *Encyclopædia Britannica*, 2003–5).

Index

A

abiogenic, 72
Acheulean (Industry), 31, 39, 171
Adams, 77
adaptation, 26
adenosine triphosphate (ATP), 7, 17, 18, 20, 92, 99, 144, 171
aerobe, 14, 16, 18, 24, 25, 29, 93, 96, 97, 99, 130, 171
Aerobic Regime, 21, 24, 25, 26, 41, 92, 93, 95, 97, 108, 112
aerobic respiration, 13, 15–18, 21, 26, 43, 92, 97, 171
Aeroculture, 113
Age of Reason, 119, 171, 175
agrarianisation, 52
Agrian Period, 54, 55, 63, 65, 71, 79, 106, 110, 118
Agricultural Revolution, 21, 51, 52, 54, 55
agrocultural, 60 (metabolism), 65, 94, 95, 97, 99 (sociosystem), 110, 111
Agrocultural Man, 95
Agrocultural Regime, 54–56, 59, 61, 63, 94, 98, 99, 109, 110, 112, 171
Agrocultural Revolution, 114
agrocultural sociosystem, 99
agroculture, 94, 95, 113
Agro-Energy Revolution, 51, 54, 55, 62, 63, 117, 143
amoebas, 19
anaerobe, 14, 16, 18, 171
anaerobic, 17 (combustion), 18 (fermentation), 92 (phototrophs)
anaerobic respiration, 17, 18, 171
ancient sunlight, 70, 72, 95, 99, 111, 130, 144

anoxygenic photosynthesis, 9, 180
Anthropic Era, 106, 139, 143, 171
anthropocentrism, 132, 138, 139, 143, 171
anthropogenic, 58, 63, 93 (fire), 126 (energy flow), 129, 130 (forcing), 171
anthropomorphisms, 7
anthroposphere, 63, 106, 108, 111, 114, 137, 171
anthroposystem, 106, 112, 171
Arcadian, 119, 120–122, 125–129, 132, 133, 135, 136, 138–140, 142, 143, 147, 148, 172, 177
archaea, 18, 19
Archaean, 4, 15, 16, 172
Aristotle, 56
Arrow of Time, 101
arto-information, 103
artosphere, 103, 105, 137, 172
Atkins, 150
Atlantis, 120
attractor state, 89, 157, 158, 161, 172
Aunger, 105
autocatalysis, 89, 99, 156, 158, 172
autotrophy, 11, 24, 172

B

Bacon, 58, 59, 119, 125
bacteria, 18, 19, 160
Bada, 4
Baumgärtner, 133
Bell, 78
Berners-Lee, 75
Big Bang, 98

biological engine, 71
biological evolution, 23
biosphere, 93, 172
black smoker, 7, 91
blue-green, 8–16, 18, 20, 22, 67, 88, 91–93, 136, 172, 174
Boltzmann, 162
Born-Haber, 74
brain-to-body ratio, 32
Burney, 84

C

Cambrian Explosion, 23, 24, 100
Cambrian Period, 172
Carbian Explosion, 71, 76, 82–84, 96, 111, 118, 132, 143, 148
Carbian Period, 70, 71, 79–81, 82, 129
Carbo-agriculture, 74
Carbocultural Man, 95
Carbocultural Regime, 70, 71, 95, 98, 99, 110–112, 121, 122, 129, 136, 172
Carbocultural Revolution, 114
carbocultural sociosystem, 99
carboculture, 76, 78, 80, 84, 96, 99, 113, 148
Carbo-Energy Revolution, 90, 96, 117, 143
carbohydrate crude, 16
carbon cycle, 11, 15
Carbon Valley, 121, 124, 145, 147
carbon-fixation, 8
carrying capacity, 62, 83, 117, 118, 139, 173
cell designs, 19
cell metabolism, 45, 46, 90, 144
Cenozoic Era, 100, 173
Chaisson, 94, 98, 154, 159, 160, 162
chemical soup, 91
chemical vortex, 155
chemolithotrophic, 8
chemotopes, 5
chemotrophy, 173
chlorophyll, 9, 11, 173, 174
chloroplast, 20, 21, 25, 173–175

CityPlex, 146, 147
Clausius, 151
clean fossil, 124, 173
climate change, 26, 50, 52, 53, 131, 132, 136, 144
closed-loop economy, 132, 135, 143, 144
CoalPlex, 147
co-evolution, 26, 29, 32, 35, 45, 46, 52, 58, 80, 104, 106, 109, 112, 114, 122, 157, 173
Cohen, 57
competition, 26
complex adaptive system (CAS), 101, 156, 173, 178, 179
complexity, 61, 98, 100, 101, 103, 138, 149, 155, 173
computer model, 88
conceptual model, 88
consilience, xiv, 80, 138, 139, 149, 162, 173
conversion path, 92, 93, 95–97, 135, 161
Copernican Revolution, 57, 138
Copernicus, 57
Crosby, 56, 57
cultural selection, 105, 137
cultural signal, 118
cytochrome oxidase, 15
cytoplasm, 19, 174
cytoskeleton, 19, 174

D

d'Alembert, 58
Daly, 135, 139
Darby, 67
Darwin, xviii, xix, xx, xxi, 18, 101, 105, 137, 160, 176
Dawkins, 100, 101, 104, 105, 137
Day, 77
de Duve, 5, 11, 14
de Rosnay, 118, 120
de Sauserre, 77
de Vries, 54
Deacon, 32, 33, 34, 35, 36, 37

INDEX

Descartes, 58, 59
Diderot, 58
diffusion, 107, 110, 138, 144, 174
dinosaurs, 29
discovery, 107, 110, 138, 144, 174
dissipative path(way), 87, 94, 151, 158
DNA, 6, 19, 20, 46, 80, 91, 92, 105, 120, 157
Douthwaite, 120
Drake, 72
driving force (see force, driving)
Dugatkin, 104
Dunbar, 36, 37
Dürr, 126
dynamic steady state, 89, 101, 161, 174

E

ecocentric, 139, 143
Ecocultural Regime, 126
eco-energy vortex, 158
ecological economics, 133, 135, 141, 174
ecological footprint (see footprint)
ecological regime, 44
ecosphere, 102, 118, 174
ecosystem, 19, 23, 29, 83, 89, 93, 100, 130, 133, 142, 157, 159–162, 174
Edison, 73
Egyptian Old Kingdom, 61, 100
Eigen, 7, 98, 103, 156
Einstein, 77, 78
Eisenhower, 78
ElBaradei, 131
emergence (emergent property), 37, 42, 43, 48, 94–96, 119, 150, 153, 154, 156, 160, 175
endosymbiosis, 20, 175
energetics, xviii, xxi, 17, 37, 43, 70, 111, 129, 133, 160–162, 175
energy culture, 117
energy economy, xvii, 24, 37, 42, 43, 69, 70, 81, 89, 91, 94, 161
energy era, 8, 44, 89, 142

energy flow, 89, 93, 96, 97, 99, 101, 115, 148, 149, 155, 160, 162
energy gradient, 89, 92, 93, 96, 97, 159, 160, 162, 175
energy mastery, 43, 142
energy period, 175
energy quality, 87, 93, 150–152, 155, 159
energy regime, 8, 97, 122, 143, 175
energy revolution, 63, 69, 70, 103, 117, 136, 139
energy sink (a high entropy state), 89, 92, 94, 97, 99, 161, 175
energy source (a low entropy state), 8, 12, 89–91, 94, 97, 99, 127, 143, 161, 175, 180
energy system, xi, 16, 37, 89, 95, 122, 127, 161
Energy Time Scale, 8, 15, 21, 143
energy vortex, 158, 160, 161
energy-dissipating paths, 89, 94, 151, 158, 159
energy-dissipating structure, 88, 89, 91, 93–95, 97, 99, 103, 105, 115, 119, 128, 132, 134, 135, 153, 155–157, 158, 160, 161, 175
energy-driven organisation, 102, 153, 155
eucarya, 18, 19
eukaryosis, 20
eukaryotes, 6, 19–24, 26, 27, 175
eukaryotic, 18, 22, 93, 100
Eukaryotical Revolution, 21, 22
evolutionary energetics, 24, 37, 38, 51, 69, 99, 113, 128, 130, 176
exergy, 100

F

far from (thermodynamic) equilibrium, 88, 89, 91–93, 100, 111, 128, 139, 153–155, 157, 161
fermentation, 17, 18, 92, 171, 176, 181
Fermi, 77
Fertile Crescent, 53, 176

fire master, 38, 83, 94, 95, 130, 142
fire mastery, 39, 43–45, 48, 49, 84, 93, 94, 108
fire metabolism, 46
Fire Regime, 176
Fischer-Kowalski, 45
flagella, 19
flow of energy, xxi, xxiv, 9, 45, 46, 60, 76, 81, 91, 93–103, 106, 111, 115, 124, 126, 128, 130, 133, 141, 143–148, 153, 155, 160–162, 176, 181
footprint (ecological; human), 61, 63, 81, 83, 84, 118, 119, 127, 139, 145, 176
force, driving, xxi, 89, 90, 91, 94, 95, 102, 155, 160, 176
force, shaping, xxi, 89, 91, 94, 95, 102, 103, 106, 155, 160, 176
Ford Doolittle, 6
Fortey, 6, 7, 13, 20, 25
Fourneyron, 76
Francis, 76
Freiberger, 75
Fritts, 77

G

Gaia, 161, 176
Galilei, 57
Garden of Eden, 119
genetic evolution, 101
genetic symbiosis, 139
genome, 20, 102, 103, 176
genus *Homo*, xxii, 27, 31, 44, 48, 62, 63, 83, 84, 93, 95, 106, 117, 142, 171, 177
geologic aeon, 15
Georgescu-Roegen, 133–135
Gesner, 72
Giardia, 19
global warming, 152, 176
Goudsblom, 39, 40, 44, 45, 54, 62, 84, 93, 110, 135

Great Rivers, 53
Green Valley, 122, 125–128, 132, 143, 147
greenhouse effect, 152, 177
Greenpeace, 125

H

Hadean, 4, 177
Hahn, 77
Halbane, 5
Helian Period, 143
Heliocultural Regime, 142, 143
heliocultural revolution, 139
helioculture, 139, 144, 148
Helio-Energy Revolution, 117, 139, 142, 143
Helioic Era, 143
heterotroph, 11, 14, 16, 130, 172, 177
heterotrophy, 24, 177
Holocene, 51, 52
Hominidae, 30
Hominids, 29, 36
Homo ecologicus, 120, 138, 141–143, 172
Homo economicus, 120, 138, 141–143, 177
Homo energeticus, 139, 141–143
Homo erectus, 40, 83
Homo ergaster, 31, 38, 40, 46
Homo heidelbergensis, 31–33, 38, 40, 42, 46, 83
Homo sapiens, 42, 46–49, 51, 59, 63, 71, 83, 84, 109, 117, 118–121, 137, 143, 148, 182
Homo sapiens carbonius, 71, 74, 78, 79, 81–84, 95, 96, 97, 129, 130, 136
Homo sapiens' societal metabolism, 56
human acting, 106, 112, 113, 114, 115, 138
human advantage, 32, 35, 37, 38
human capacity, 106, 112–114, 138
human civilisation, xiii, xxiii, 38, 52, 63, 93, 98, 100, 135
Human Development Index, 119
human evolution, 38, 42, 48

INDEX

human knowing, 106, 112–114, 138
human living, 106, 112–114, 138
Hüttler, 45
Hydrocarbon Man, 65, 71, 82, 83, 84
hydrothermal vent, 8
hyperthermophile, 6–8, 11, 89, 91, 92, 100, 177
hyperthermophile Eden, 6
hyperthermophile Noah, 6

I

Ice Age, 50, 51
Imperial, 119, 120, 121–126, 128, 131, 132, 138, 139, 140, 142, 143, 148, 172, 177, 182
Industrial Age, 76, 113
Industrial Ecology, 138, 147
Industrial Revolution, xxiii, 21, 69, 71, 103, 111, 178
industrial symbiosis, 69, 76, 146
innovation, ix, xxii, 66, 105, 107, 108, 110, 111–114, 127, 131, 136, 138, 140, 141, 144, 145, 147, 148, 178
internal combustion, 15, 16, 43, 45, 71, 73, 82
International Atomic Energy Agency (IAEA), 131
International Energy Agency (IEA), 122, 178
invention, 107, 110, 138, 144, 178

J

Jablonski, 23
Jansen, 57
Jørgensen, 100

K

Kaku, 120
Kauffman, 89, 101, 156–159, 161
Kay, 158, 159

Klein, 30–32, 40, 48
Knoll, 15, 16, 19–21
Kondratev, 111

L

Landes, 103
Lavoisier, 58
Law(s) of Thermodynamics, 87, 132, 150, 155, 158, 182
Lenton, 14, 15, 16
Levantine Corridor, 53, 178
Levin, 118
Lewin, 32
Living Planet Index, 83, 119
Lomborg, 118, 121
Lovelock, 161
Low Countries, 65
Lower Paleolithic, 39
Lowland Classic Maya, 61, 100
Lunine, 153

M

macroscope, 118
macroscopical sciences, 138
Macroscopical Signal, 90, 117, 119, 121, 128, 132, 135, 138, 148, 178
Malthus, 62
Man the Fire Master, 42, 71
Man the Solar Farmer, 71, 83
Man the Talker, 32
Man the Toolmaker, 32, 106, 108
Marchant, 54
Margulis, 6, 20, 24, 149, 155
Mars, 11
Marten, 120
measurable reality, 56
mechanical civilisation, 69
meme, 103–106, 109, 110, 115, 137, 173, 178
memetic(s), 104–107, 137, 139, 143
memetic symbiosis, 138, 139
Mesopotamian Empire, 61, 100

Mesozoic, 100, 178
metabolic revolution, 72
methanotrophs, 15
Microbial Symbiosis, 18
Middle Ages, 56, 70
mindsphere, 106
mitochondria, 20, 21, 24, 25, 45, 179
morphological evolution, 23
Morse, 74
Mouchot, 77

N

natural capital, 133, 141
natural selection, xix, xx, xxi, 16, 26, 91, 101, 102, 137, 179
near-closed-loop, 130, 134
neoclassical economics, 131–133, 177, 179
New Age, 138
Newcomen, 67
Nisbet, 6, 21
Noah, 6
noösphere, 105, 106, 137, 179
Nowak, 36
nuclear, 3, 4, 77, 78, 96–98, 115, 122, 131
Nuclear Valley, 124, 125, 143, 147
Nucleocultural Regime, 123

O

Oldowan Industry, 31, 179
Oresme, 57
organelle, 19, 20, 25, 173–175, 179–181
Oxian Period, 21, 43, 90
Oxo-Energy Revolution, 13, 21, 90, 96
oxygen crisis, 14, 27, 92, 93, 136
oxygen holocaust, 16
oxygen oases, 15, 17
oxygen revolution, 16
oxygen-enriched air, 14
oxygenic photosynthesis, 9, 11, 15, 17, 18, 20, 21, 26, 180

P

Palaeolithic Period, 171, 180
paleo-record, 14
paleo-signs, 47
Paleozoic, 100
Panspermea, 5
Pearson, 78
Perlès, 46
Permian, 100
phlogiston, 58
Photian Period, 11, 12, 15, 18, 43, 67, 90
photoautotrophy, 11, 24
Photo-Energy Revolution, 3, 10, 12, 90, 96
Photoic Era, 11, 43, 90
Photosphere, 3, 180
photosynthesis, 8, 9, 15, 17, 20, 24, 43, 87, 97, 143, 152, 153, 173, 180
photosynthesis technology, 10, 11
photosynthesiser, 8, 14, 16, 21, 94, 99, 101, 126, 130
photosynthetic, 9, 10, 13, 14, 16–18, 21, 22, 24, 25, 27, 43, 87, 91, 92, 97, 129, 172, 174
phototrophic ecosystem, 99
Phototrophic Regime, 10, 12, 16, 91, 97, 136
phototrophy, 11, 180
Pimm, 83
Plato, 56
Pleistocene, 50–52
Ponting, 65
Prigogine, 154, 157
primate, 30, 34, 36, 40, 51, 103, 177
prokaryote, 19, 22, 181
Prometheus, 40
Proterozoic, 15, 22, 181
protocells, 157
proto-ecosystems, 89, 158
protozoan, 22, 181
Pyne, 39, 40, 42, 44, 45, 48, 50, 62
Pyrian Period, 50, 51, 62, 71, 83, 90, 118
Pyrocultural Revolution, 114

INDEX

Pyrocultural Regime, 43, 44, 50, 55, 63, 90, 93, 96, 99, 108–110, 112, 181
pyrocultural sociosystem, 99
pyroculture, 93–95, 113, 117, 121
Pyro-Energy Revolution, 29, 44, 62, 90, 96, 106, 117, 143
Pyroic Era, 44, 50, 51, 55, 71, 84, 90, 142
pyrotechnological innovations, 43

Q

Quantificational Signal, 56, 57, 90, 110, 111, 114, 117, 181

R

red bed, 13
reductionism, 59, 80, 139, 181
reductionistic science, 112, 113, 132
Rees, 118
Reijnders, 126
Renaissance, 61
Robbins, 73
Robo sapiens, 120
Roman Empire, 61, 100
Ronen, 38, 39
RuralPlexes, 147

S

Sagan, 6, 20, 24, 149, 155
Savery, 67
Schellnhuber, 57
Schneider, 158, 159
Schumpeter, 111, 112
science schism, 59, 181
scientific reductionism, 138
Scientific Revolution, 57
selection pressure, 30, 34, 94, 95
self-organisation, 101, 102, 105, 154, 155, 181

shaping force (see force, shaping)
Sieferle, 68
Sleep, 6, 21
Smil, 59, 69, 72, 81, 121, 122, 141
Smith, 102, 131, 132, 177
social evolution, 35, 37, 38, 42, 43, 63, 109
societal metabolism, 45, 46, 65, 117
socio-chemical metabolism, 93
socio-metabolic complexes, 145, 147
socio-metabolism, 78, 81, 129, 143–145, 181
sociosystem, 89, 97, 99, 102, 103, 127–129, 131, 132, 135, 138, 139, 142, 160–162
socio-technological development, 106, 107, 113–115, 125, 130, 143
socio-technological progress, 124
socio-technological revolution, 113, 114, 143
Soddy, 70
solar farmer, 51, 109, 130
sound languages, 36
source-sink system, 92, 93, 96
source-to-sink, 97, 161
Sparnaay, 57
Staircase of Evolutionary Progress, 101
Staircase of Socio-Technological Development, 107, 108, 111–115, 147
steady state, 89, 101, 106, 111, 122, 135, 152, 161, 172, 181
steady state economy, 135
steam engine complex, 69
steps of socio-technological development, 108
Stone Age, 46, 48, 180
Strassmann, 77
sugar crude, 10, 18
sulphur bacteria, 7
Sumer, 60
Sun Valley, 143, 145, 147

survival of the fittest, xix, xx, xxii, 137
sustainable development, 115, 118, 128, 129, 131, 138, 139, 143
Swaine, 75
Symbian Man, 136, 139, 142–144, 148
symbiocentric, 139, 143
symbiogenesis, 175, 182
symbiosis, 18, 20, 27, 136, 138, 139, 146, 181, 182
Symbiotic Planet, 136
symbolic reference, 34, 35, 37, 39
symbolic thinking, 109, 112, 142
Symbolisational Signal, 46, 48, 90, 109, 110, 114, 117, 182
symbolization, 33
syntax, 37

T

Tainter, 60, 61, 98
technosphere, 103, 182
thermodynamic equilibrium, 88, 89, 91, 128, 153, 161
Thermoic Era, 8, 11, 89, 90
Thermophilic Regime, 8, 89, 90, 96, 97, 99
tree-dwellers, 30

U

United Nations, 82, 96, 126, 129

V

Valéry, 111
van Klinken, 126
Verloop, 113
Vesalius, 57
von Helmholtz, 158
von Siemens, 77

W

Ward, 29
Watt, 68, 69
Wills, 4
Wilson, 58, 79, 80, 138, 162
wind-water machine, 9, 10, 17
World Bank, 82
World Business Council for Sustainable Development, 129
World Resources Institute, 129
World Wide Fund For Nature, 83, 96, 118
Worldwatch Institute, 82
Worster, 119, 120
Wright, 82

Z

Ziegler, 126
Zimmer, 23, 29, 37